Containment in the Pharmaceutical Industry

DRUGS AND THE PHARMACEUTICAL SCIENCES

Executive Editor

James Swarbrick
AAI, Inc.
Wilmington, North Carolina

Advisory Board

DRUGS AND THE PHARMACEUTICAL SCIENCES

A Series of Textbooks and Monographs

Containment in the Pharmaceutical Industry

edited by

James P. Wood

Eli Lilly and Company
Indianapolis, Indiana

MARCEL DEKKER, INC.

NEW YORK • BASEL

Great care has been taken to maintain the accuracy of the information contained in this volume. However, neither Marcel Dekker, Inc., nor the editors nor authors can be held responsible for errors or for any consequences arising from the use of the information contained herein.

ISBN: 0-8247-0397-9

This book is printed on acid-free paper.

Headquarters
Marcel Dekker, Inc.
270 Madison Avenue, New York, NY 10016
tel: 212-696-9000; fax: 212-685-4540

Eastern Hemisphere Distribution
Marcel Dekker AG
Hutgasse 4, Postfach 812, CH-4001 Basel, Switzerland
tel: 41-61-261-8482; fax: 41-61-261-8896

World Wide Web
http://www.dekker.com

The publisher offers discounts on this book when ordered in bulk quantities. For more information, write to Special Sales/Professional Marketing at the headquarters address above.

Current printing (last digit):
10 9 8 7 6 5 4 3

PRINTED IN THE UNITED STATES OF AMERICA

Preface

The pharmaceutical industry, like the rest of the healthcare sector, has been changing the way it does business. Companies are reevaluating just how far outside their comfort zones they are willing to operate, and this is driving change at rates beyond what the industry has traditionally seen. In terms of technology, one of the fallouts of this is the relatively quick evolution of containment approaches for compounds and processes in many pharmaceutical operations. That in itself is an exciting and promising development, but it creates the challenge of keeping whatever is written from becoming obsolete by the time it is published. Keeping material current is problematic, especially given the time involved in publishing a book presenting full coverage.

An additional challenge is the ambitious scope itself. Globally, the "pharmaceutical industry" is made up of a far-flung and diverse set of companies, each with its own experiences, philosophies, and policies. There is, therefore, no uniformly agreed-upon set of specific standards that cover all applications or contingencies.

However, given the above challenges, we have striven to present not only up-to-date information, but also timeless approaches and advice that should remain beneficial to the audience even after technologies have further evolved.

This book has been developed with an eye to positioning it in a somewhat different light from other publications. Other books have been written (and very well written) with a focus on specific containment-related topics, such as barrier technology, isolator applications, and aseptic isolator designs in pharmaceutical processes. Instead, this book finds its niche in taking a more holistic view of containment overall, as applied and achieved (and sometimes misapplied and unachieved) in the industry; the range of discussion will be more encompassing than what is typically seen in books on containment. And while that oftentimes can exact a price in the sacrifice of some detail, we have tried not to lose too much in depth of discussion of the main topics of interest. Often, a book that tries to be all things to all people tends to disappoint some and discourage the rest—we have tried to avoid that trap here.

Because there is no universal industry agreement on methods to contain— or, at times, even how to define the specific hazards involved—the objective of *Containment in the Pharmaceutical Industry* must be the exploration of the main elements of containment as it is currently practiced by various members of the pharmaceutical industry. That individual views may at times be at variance with one another should then come as no surprise. Specifically, this book first presents a brief historical perspective of containment in the industry, a workable definition of the term, and containment's relationship to other industries. We include this overview to provide a context in which to better understand why containment has evolved as it has from the recent past to where it is today in the industry. Along with initial definitions, a chapter focusing on some recent research of what makes powders ''dusty''in the first place and how that is being defined, quantified, and used in containment design is included. This book discusses approaches for both new and retrofitted installations, and there is discussion of ''people protection'' versus ''product protection'' and source containment's potential role in each. In addition, this book offers consultants' and engineering firms' perspectives, along with pharmaceutical firms' views.

That containment should be ''built into'' the process or system from a project's inception is one of the underlying themes the reader will find throughout this book. That being the case, it is reasonable to expect that the concepts and tenets of containment will be integrated throughout the entire lifespan of a project. A project's planning, design, construction, and start-up commissioning phases must all reflect the various containment provisions and philosophies with which the project began. Consequently, this book will be of interest to those who are responsible for the planning, design, building, and start-up activities of projects that have containment elements. Also, because the production process or facility obviously is intended to keep meeting its containment criteria beyond day one, individuals with the responsibility for ongoing operation and maintenance of contained processes have an interest in the topic as well. More and more, companies are realizing that certain forms of continuous, at-the-source containment are economically justified. Often there is also an interest in retrofitting existing processes for targeted source containment, in which case plant and operating personnel might also be interested in hearing views on the feasibility (pros and cons) of retrofitting for containment in an existing area or for an existing process.

This book is meant to be one tool among several others that you may have at your disposal, which can be useful in the quest to better contain your pharmaceutical operations. Your decision on both the degree and the approach to containing any operation needs to be based on sound business judgment. To that end, the better the decision-maker understands the operation in question—both overall and in its nuances—the more an informed selection can be made of which approach to pursue. This is key, as there is often more than one possible solution. The information gleaned from this book will make you aware of the business case for effective containment, aid you in better defining any process emissions problem you might believe you have, and acquaint you with containment challenges and approaches for many areas in the pharmaceutical industry.

This book is dedicated to the pharmaceutical industry, striving in its journey for on-going improvement and elevation; and to my father, James C. Wood, for his embarking on that journey long ago.

James P. Wood

Contents

Contributors

Melvin J. Crichton, P.E.   Engineering Tech Center, Eli Lilly and Company, Indianapolis, Indiana

Walter W. Czander   Project Management, Lockwood Greene, Augustus, Georgia

Donna S. Heidel, C.I.H.   Safety and Industrial Hygiene, World Wide, Johnson and Johnson, New Brunswick, New Jersey

Dane O. Kildsig, Ph.D.   Department of Industrial and Physical Pharmacy, Purdue University, West Lafayette, Indiana

Edwin A. Kleissler, P.E.   Kleissler Company, Lakeland, Florida

Steven M. Lloyd   Research and Technology Department, ILC Dover, Inc., Frederica, Delaware

M. Michele Moore, R.Ph.   Containment Technologies Group, Inc., Indianapolis, Indiana

Dave Palister   Extract Technologies, Somerset, New Jersey

Chetan P. Pujara, Ph.D.   Pharmaceutical and Analytical Research and Development, Abbott Laboratories, North Chicago, Illinois

Hank Rahe   Contain-Tech, Inc., Indianapolis, Indiana

Brian G. Ward, Ph.D., C.Chem., F.R.S.C.   Containment Technology/Industrial Hygiene Technology, Eli Lilly and Company, Indianapolis, Indiana

Ronald W. Wizimirski   ILC Dover, Inc., Frederica, Delaware

James P. Wood, P.E.   Containment Engineering Technology, Eli Lilly and Company, Indianapolis, Indiana

1
Why Contain? Then and Now

James P. Wood
Eli Lilly and Company, Indianapolis, Indiana

I. DEFINITION

So you've got a process that will be handling a compound that's either highly potent, highly toxic, or worrisome in some other aspect. In any event, wherever the exposed compound might wind up throughout your facility, through emissions, accidental spills, or other means, the exposure becomes a potential source of hazard for your employees, a potential source of cross-contamination for your other products, and a potential source of headaches for your project-planning team. "This process has to be contained," the project team agrees. But do they really agree? After some debate and several cross-discussions about what is meant by *containment*, different project team members around the table begin to cast about for various criteria to define the term. The realization begins to dawn that a common definition to the word *containment* might be useful. Getting back to basics for a moment, the first inclination might be to look up *containment* in the dictionary. Webster has, of course, a variety of definitions; all very correct, but all unusable for pharmaceutical purposes. (kən-'tān-mənt . . . the act, process, or means of containing; syn. restrain, check, halt, control. . . .)

As more players in the industry become more involved in containing their processes, a number of definitions for *containment* will doubtless evolve. In fact, they already have. For the purpose of this discussion, we submit an operative definition here, along with a brief discussion of the key components of containment that is in tune with the industry.

If the overall manufacturing function is thought of as consisting of three elements: product, people, and the surrounding environment, containment consists of isolating the first of these elements (product) from the other two.

1

Three points about this definition bear discussion immediately.

1. It begs the question of defining another term: *isolation*. Specifically, isolated to what degree? Contained to what level? Subsequent chapters discuss commissioning and verification strategies, air monitoring methods, and similar quantification issues. For now, suffice to say there will be an emission threshold driven by quality assurance, industrial hygiene, toxicology, or other functions, below which the process will be deemed "isolated."

2. Playing on Point 1, isolation rarely means absolute zero emissions. Even if monitoring assays result in zero-detect readings, all that can be said is that containment concentration levels are below the level of detection for that particular test. In the preceding definition, limiting product migration to the other two elements (people and the environment) to below defined thresholds set by the appropriate groups, be they Quality, Industrial Hygiene, or other, will be taken to mean the product is isolated.

3. The overall definition as stated is straightforward and purposely left in open-ended terms. It doesn't presuppose specific solutions, engineering or otherwise, prior to a full-fledged engineering analysis, but it keeps as open as possible the range of potential approaches to meet the specific design criteria for the process or compound in question.

II. HISTORY

Containment, like other aspects of pharmaceutical manufacturing, has evolved over time. Also like other aspects, this evolution hasn't always been a steady one, but a series of plateaus punctuated by relatively quick jumps. Helping this evolution has been the fact that other industries have, over time, needed effective containment as well. The nuclear industry is an obvious example. Referring to our three-element definition of containment, the surrounding environment and people elements need to be protected from the radioactive product element at hand. Likewise, the asbestos abatement industry employs extraordinary means at times to contain its product from people. Conversely, the aeronautical and electronics industries, as well as the sterile pharmaceutical industry, strive for ultra-clean processes via isolation techniques to protect their products from people and the surrounding environment. Finally, hospitals and similar operations need to isolate certain agents they handle from people and the environment for the benefit of all three of those elements. Of course, as these industries shared their experiences back and forth with one another, it became evident that what was a good idea for one wouldn't always turn out to be applicable for another, due to different sets of criteria. The point is by sharing the knowledge base, the evolution of containment techniques and philosophies was able to progress.

A. Focusing on the Pharmaceutical Industry

When the pharmaceutical industry's modern era was in its infancy, defined here as between the latter nineteenth and very early twentieth centuries, the importance of containing certain compounds and chemical agents was not as appreciated as today, just as the challenges of doing so were not as well understood. The cost of the product was driven by other aspects of production, e.g., the cost of raw materials, equipment, facilities, people. If a compound happened to be particularly noxious, as demonstrated by operators' eyes watering or other noticeable, acute reactions, some form of mask or general remedy was often put in place after the fact.

After this era, firms came to depend more and more on the rooms housing the process to be the primary containment means for any emissions from the process that might occur. With some exceptions, this general philosophy held sway for quite a long time. In fact, to a large degree, a significant part of the industry retains this mind-set today. The considerable cost of constructing "contained facilities" adds to the other cost components of production. The room, then, was the primary containment device. If a compound was being handled that was particularly troubling, local exhaust (theoretically, but rarely, located effectively next to the source of the emission) was employed. The room still served as the primary safety shield to the outside world. It's important to note that during this time, these measures were generally not thought of in terms of returns on investment. Rather, they were typically safety driven, and at times also quality driven. They were seen as a necessary cost of doing business, with no immediate or definable financial payback.

As time moved on, the toxicological sciences continued to acquire in-depth understanding regarding potential subtler and longer-term physiological effects of chemical agents. Compound potencies rose. The design and manufacturing of personal protective equipment (PPE) also improved, and PPE was utilized to greater and greater extents. Differentiation of PPE grades and types became greater and more defined. Again, the prime driver for all of this was a heightened awareness of the hazards involved and, therefore, the increasing efforts required to achieve safety thresholds. Little or no immediate investment return was factored into the analysis. For a period of time, this became a benchmark of "containing compounds." In effect, the operators were carrying their own personal safe environment, their PPE, along with them. Relating to our earlier definition of containment, their breathing masks and suits were isolating the people element, while the "contained room" protected the environment outside the room. Both elements were individually and separately isolated, after a fashion, from the product element.

Of course with this development, the containment cost of production increased once again. PPE and room isolation carry their own costs in initial capital outlay, as well as owning and operating costs.

Initial costs included, of course, purchase of the suits, but also (if containment were to be maintained with integrity throughout) the requisite gowning and degowning facilities; utilities and drainage for decontamination showers, with all attendant isolation, airlocks, isolated HVAC, and exhaust systems (such mechanical systems located also in contained areas for safe maintenance and repair access); pressurization controls; special filtration and filter-change hardware and procedures; room decontamination equipment and protocol; contaminated suit storage provisions; isolated clean suit storage provisions; and, of course, the additional expensive square footage for these areas to exist within the production building.

Ongoing costs included the energy cost of throwing away temperature- and humidity-conditioned exhaust air; additional exhaust fan and HEPA filter testing and preventive maintenance from the areas; regular (often daily) decontamination of the gown/degown rooms along with production areas; the cost of periodic suit replacements; and the less tangible but potentially significant cost of increased cycle times. To get a grasp on this last point, consider that in some operations each operator might leave the production area a minimum of three times per shift, a lunch and two breaks. Each of those instances has associated with it the time required to go through multistep decontamination, degowning, and then regowning (potentially with a new clean suit) throughout the day. This can occur three times each shift, each day, for each worker.

Now enter what appears to be on its way to becoming the next plateau in containment, which can be roughly referred to as ''contain-at-source.'' I liken this concept to Pandora's box. While the lid remains closed, the surrounding world gets along just fine. But once the lid is opened, all the sins and evils contained within escape out, to the eternal woe of everyone. The rationale for containing at the source is that simple. If someone could just keep Pandora from opening the box in the first place, there wouldn't be any need of going out into the wide world to clean up all the released mess. It's easier to manage the compound, or product element in our definition, while it's restricted to a smaller volume, than after it's spread out over an exponentially larger space, including wall surfaces, equipment, exteriors of operators' breathing suits to be tracked out, ad infinitum.

Given that the history of each new containment benchmark, at least at first glance, has added to the direct cost of production, it's not surprising that production managers and engineers alike might have an off-the-cuff reaction along the lines of ''here comes another cost increase.'' However, as members of the industry began putting pencil to paper, some interesting developments began to arise. While the room-containment approach isolated (sometimes) the outer environmental element from the product and people, and PPE isolated the people element from the product, directly containing the product element to begin with minimized the need for these other remedies, driving their costs down to less significant amounts.

III. THE BUSINESS CASE

A. In Worker Health

In the previous section, the direct consequenes and costs in not containing at the source but relying on secondary, or remedial, containment were discussed: extra square footage, decontamination requirements, increased plant cycle times. Realize, however, that the costs mentioned up to this point have been driven by worker health and safety concerns. With just that piece of the analysis, performed in sufficient detail and covering all true costs, a cost-avoidance/payback analysis can be at times surprising. For example, what would be the cost savings in PPE and related decontamination alone, if the process were contained instead to a shiftsleeve room environment level? But, of course, as vital as employee health is, the pharmaceutical industry has additional masters to serve. The potential for cost avoidance in meeting these other demands boosts the business case for good containment even further.

B. In Product Quality Assurance/Cross-Contamination

How much does each swab sample for a room or hallway cost your company? Of course, purchase of the sample container and filter or medium is needed and is probably the most definable piece of the total. However, other costs are incurred as well. These include such things as the cost of labor in collecting the sample, additional quality control "blanks" and "spiked" samples for the analytical lab, transport to the analytical lab, lab time, equipment, personnel capacity supporting that incremental demand for sampling analysis, and results interpretation (someone has to write the report). With large and regular sampling programs, these incremental sampling demands add up to real lab-capacity demand, facilities, and personnel requirements.

And, of course, don't forget that all this doesn't happen in an hour or two. During the days or weeks between the cleaning, sampling, lab assays, and reporting back to the user confirming that "all is clean," the facility's switchover time clock continues to tick. The facility at this point is nothing but "idle plant" as far as production is concerned. So what's the internal cost for this? Taking the time to establish the true, total cost to a company is rarely straightforward, but almost guarantees to be eye opening. Now, if the compound in question, or product element of our containment triad, could be more effectively trapped within its source before room contamination occurs, the cost impact to the quality sampling programs becomes evident. Indeed, if an incremental improvement in containment served only to contain certain operations that were emission sources previously, at least the rooms housing such operations could eliminate or reduce the frequency of such sampling requirements, affecting the previously mentioned costs.

C. In Environmental Issues

Let's explore a few of the cost elements in assuring ourselves and the public that the air and water streams emanating from the plant are not hazardous.

1. Water

First, there are liquid-waste streams. When a room or suite of rooms is contaminated, oftentimes the fix is to spray down those entire areas in one form or another. Assuming the spray down is successful, the contaminants will now be, if all goes well, contained in the water stream. Now the plant personnel need to figure out what to do with that large amount of water, since they have in effect succeeded in taking, for example, a 100-to-1000g powder-contamination problem and transforming it into a 100-to-1000 pound water-contamination problem. Depending on the specific contaminant involved, maybe this is one of the more workable solutions. But with the advent of higher and higher potency compounds, cytotoxins, endocrine disrupters, etc., being handled in the industry, often the only solution the user feels comfortable with may be along the lines of collecting all waste water in a separate system and hiring a third party to regularly pump into tankers for remote incineration.

 Even assuming for the sake of discussion that this approach is the optimal one, the variable most directly affecting the cost is the quantity of contaminated water that is generated. If that quantity can be cut (through, for example, improved containment reducing the need for area wash downs or a reduced amount of water per washing), the initial capital cost of a collection system will be lower due to smaller system capacity requirements. And ongoing costs will also be reduced by the lower frequency of tanking off-site for disposal. Of course, if the method of waste-water treatment is something other than off-site incineration, the preceding observation is still valid, as long as waste-water quantity is the key variable in the costs incurred.

2. Air

If the room air exhaust in your process areas works as effectively as it's supposed to, what does that mean? It means you've successfully taken a contamination problem previously confined to a 5,000 cubic foot room and, by design, engineered its potential escape into the open outdoors' environment. To keep that from happening, of course, companies typically resort to various air-treatment hardware, such as dust-collection systems, multiple high-efficiency filtration, scrubbers, and such. Properly designed, installed, and maintained, this approach is very effective for a variety of pharmaceutical applications. Each of these devices also incurs a cost, however, both in initial capital (purchase cost and extra building square-footage requirements), as well as ongoing costs such as maintenance and repair; filtration checks; and increased energy for extra fan horsepower to overcome system pressure drops. And this doesn't include any cost to be sure that maintenance of this equipment is performed in a contained fashion, since such equipment will be internally contaminated whenever the maintenance shops want to open it up to repair or change anything.

Again, there are always applications where it makes sense to employ an exhaust approach. Because of the costs, don't automatically assume it is the approach of choice for all circumstances. As the effectiveness of source containment on the production floor increases, there will be less need for this type of secondary ventilation-based containment. The result will be downsized air systems that are needed to begin with, a slower loading of filters during operations, decreased labor time in maintaining filters, and reduction in the other cost categories mentioned.

IV. SUMMARY

All these elements—safer processes for employees; higher assurance of no cross-contamination for the Quality Assurance department; smaller volumes of solid waste (suits, filters, and other similar materials); waste water; and exhaust air for the environmental plant to deal with—all affect the true bottom-line costs of production. To the extent that containing at the source decreases these costs, keeping Pandora's box clamped shut is just good business for the industry.

2

Containment and Good Manufacturing Practices

Melvin J. Crichton

Eli Lilly and Company, Indianapolis, Indiana

EDITOR'S NOTE

The ISPE Baseline Facility Guides are being developed to help facility designers gain a more common understanding of current Good Manufacturing Practice (GMP) expectations, while applying good engineering practices to achieve optimal utilization of capital. The International Society of Pharmaceutical Engineering (ISPE) and the Food and Drug Administration (FDA) teamed together to drive the creation of the guidelines, and they reflect a broad consensus across the industry as to engineering approaches to new facilities. This project was promoted and funded by some of the key members in the industry, including:

Alcon Laboratories
Bayer Corp.
Boehringer Ingelheim
Bristol Myers Squibb Co.
Eli Lilly and Company
Glaxo Wellcome Inc.
Hoffmann-La Roche Inc.
Merck & Co.
Pfizer Inc.
Pharmacia & Upjohn Inc.
Wyeth-Ayerst Laboratories
Zeneca Pharmaceuticals
Zenith Goldline Pharmaceuticals

These and other firms—pharmaceutical manufacturing, architectural and engineering, and equipment manufacturers—also participated in the actual development of the guidelines.

9

This body of documents will, when completed, consist of ten major guidelines pertaining to different critical aspects of pharmaceutical manufacturing. Three of the guidelines have been written and published already. They are the *Bulk Chemical Manufacturing Facilities* guide, the *Oral Solid Dosage Facilities* guide, and the *Sterile Manufacturing Facilities* guide.

Mr. Crichton's role as a member of the ISPE Technical Documents Steering Committee has been to facilitate the teams creating the guides, as well as editing and authoring sections for the guides already published. His work has also included creation of the ISPE Biotech Facilities Guide and acting as ISPE's representative to the United States Technical Advisory Group (US-TAG) for ISO TC 209, the new international clean room standard.

I. INTRODUCTION

The term GMP, also known as current Good Manufacturing Practice (cGMP), takes its roots in the Federal Regulations, CFR 210 and CFR 211. The FDA elaborated on the GMPs for the processing of sterile pharmaceuticals with the *Aseptic Processing Guidelines* of 1987 and 1992. While nothing in these GMP documents mentions containment devices, work is underway to address *barriers* (containment devices for aseptically processed products) in the next version of the *Aseptic Processing Guidelines*. However, whether an aseptic process is open or contained, there is an expectation that the product will be protected from contamination present in the general environment and from the people who work with it. To this end, the GMPs have one underlying basis—protect the product. If the product is a parenteral (to be injected into the patient) the key requirements for the process environment are control of contaminants, both inert and biological.

Other GMPs have been more specific. For example, the European Commission (EC) GMPs are prescriptive in how a manufacturer must address certain operations. Since manufacturers often produce drugs for multiple markets from one facility, it would be ideal to have just one set of GMPs for the entire world, but this has not yet happened. Inspectors from various countries have had different expectations of the same facility, and thus the facility has had to satisfy a number of different expectations. This had led to a "design for worst case" approach for sterile-product manufacturing facilities, and often for nonsterile-product facilities.

Sterile product, processed aseptically, usually requires positive airflow from the product toward the room and its personnel. The surrounding room has to be clean enough to assure that occupants and the room can not contribute significant bacterial and particulate contamination. If a sterile product is relatively harmless to personnel (usually in a liquid form), this positive airflow causes no undue health concerns for the operators. But increasingly we have seen sterile products that have some toxicity to operators. Airborne concentrations of these products can lead to increased Occupational Health and Safety monitoring, and environmental emission concerns. These lead to many of the cost issues, mentioned elsewhere in this book, in trying to meet seemingly conflicting requirements.

II. HOW WE INTERPRET THE GMPs

Because of the nonprescriptive wording in the U.S. GMPs, many companies have applied their own interpretation of GMP expectations. Sometimes these interpretations are driven by actual experiences, good and bad. Sometimes, but not as often as desired, product-protection designs are driven by data gathered during product development.

To say that sterile product manufacturing is expensive is to make an understatement. The cost of monitoring product and room environment is high. As companies began to "one up" each other, the cost of new facilities began to spiral upward over time. Often a company would learn of features that other pharmaceutical companies had incorporated into their pharmaceutical plant designs and, fearing that regulatory requirements would rise by the time their own facility was completed, embellished their plant designs, often with no noticeable benefit to product protection. This led to a "ratcheting" effect on facility costs, both capital and operational. These costs added to product cost (COPS), as well as reducing the amount of funds available for more meaningful investment like product research and development.

Over the years, various engineers, quality experts, and sometimes inspectors, have expressed viewpoints on facility design. For example, the *American Society of Heating, Refrigeration, and Air Conditioning Engineers (ASHRAE) Applications Handbook* discusses heating, ventilation, and air conditioning (HVAC) design for sterile- and nonsterile-product manufacturing, but does not delve deeply into layout, architectural finishes, processes, water systems, or personnel activities. In general, there was no single place for a designer to collect all the information needed to assure that a facility would pass inspection. Lacking comprehensive information, the cost spiral continued.

In order to curtail this escalation of facility costs, and at the same time help facility designers understand current GMP expectations, the International Society for Phamaceutical Engineering (ISPE) teamed with the FDA to create a series of *Baseline Facility Guides*. These guides will cover different types of pharmaceutical-product manufacturing facilities, and are not to be construed as being GMPs. Likewise, they are not intended to be the only workable approach to facility design. In the foreword of each published guide is a letter from the FDA stating this.

The process of creating a guide, although time-consuming, is relatively simple. Engineers and quality people from the pharmaceutical industry collectively discuss their interpretations of the GMPs for various processes. Then, when they have created a position, they discuss it with key FDA personnel. Once a concept is written into a draft, it undergoes extensive scrutiny by the remainder of the industry and FDA personnel "wordsmith" the document into its final form. Often many iterations are required to create a document that means something to a designer, yet does not create a revision of the agency's GMP expectations.

The *ISPE Baseline Guide* program encompasses ten guides, two of them will be "horizontal" and guides that cover all types of facilities: Water and Steam Systems and Commissioning and Qualification. The remaining guides are planned to be "vertical" guides, covering different types of facilities: Bulk Chemical Manufacturing Facilities, Oral Solid Dosage Facilities, Sterile Manufacturing Facilities, Biotech Facilities, Product Development Laboratories, Medical Devices, Oral Liquids and Aerosols, and Packaging and Warehousing. As of 1999, three facility guides have been published: *Bulk Chemical Manufacturing Facilities, Oral Solid Dosage Facilities*, and *Sterile Manufacturing Facilities*.

Facility guides are organized basically the same way, with chapters covering:

Regulatory Concepts
Product and Process Considerations
Architecture and Layout
HVAC
Process Utilities and Support Utilities
Electrical Systems
Other Regulations Besides GMPs
Instrumentation and Controls
Commissioning and Qualification

III. BULK PHARMACEUTICAL CHEMICALS

The first *ISPE Baseline Guide* (1) covers the manufacture of bulk pharmaceutical chemicals (BPC) (i.e., small molecule drugs). It was created by teams of engineers from ten pharmaceutical manufacturers and four engineering and equipment firms, in cooperation with FDA field personnel. Drafts were commented upon by ISPE membership from several dozen companies. It was published in 1996.

The *BPC Baseline Guide* clearly states that the manufacturer must know certain things about the product and processes in order to design the lowest-cost facility:

1. What **critical parameters** affect the product? For example, in what steps is the product temperature sensitive or moisture sensitive? What are the acceptable limits (**acceptance criteria**) or these parameters to assure the product is not adulterated?
2. What is the product's **impurity profile**? What chemical processing steps can tolerate what type of contamination? What critical steps require tight control of potential contaminants? What are the known contamination limits for each step? In what process' steps can product chemistry be affected adversely?
3. Which steps are open to the **room environment**? What are the potential contaminants in that environment? Are these steps "critical steps" that will alter the product's characteristics if not controlled?
4. **How many products** will be processed in the facility? Are they processed simultaneously or campaigned? What kind of residue do they leave in equipment? Can other products in the facility adversely affect a product?

The manufacturer needs to know as much about the product as possible, in order to design a process that will meet the final bulk product's requirements.

In most chemical processes, some contamination is expected, but the processes are very robust and are designed to remove the contamination. Toward the end of the bulk manufacturing process, the product is expected to meet the same purity levels as in the initial steps of finished product formulation in the secondary manufacturing pharmaceutical plant.

Because many liquid-chemical processing steps require solvents, acids, or bases to synthesize the product, open processes drive the facility toward high-air changes of once-through ventilation and reliance on spot-capture ventilation. Flammability issues raise the concerns of insurers, and threshold limit values (TLVs) for the solvents used place a strong burden to protect the operator. Beside creating personnel and property loss issues, high ventilation rates increase capital and operating costs.

In its final bulk form, the product often is filtered and dried to a powder. Because the final product is undiluted by excipients (which are added in the pharmaceutical plant) it can be very potent, requiring additional personnel protection. This is where containment enters the equation.

The *BPC Baseline Guide* states that many GMP, worker safety, and flammability issues can be eliminated by designing closed processes. If a process can be proven closed, facility GMP issues are of much lesser impact than for open processes. Although in our discussions we speculated that a closed bulk process can be run "in the parking lot," in reality a clean well-laid out facility is all that is required. Also, capital and operating costs for high ventilation rates often can be reduced.

The closing or containing of bulk processes shifts the GMP burden from the facility designer to the process designer. Issues such as vessel charging (often from bags or drums) take on greater significance. Sampling of vessel or dryer contents creates a challenge. The method of cleaning equipment, often quite large, requires validated cleaning processes, leading to a growth in the Clean-In-Place industry. Although a final bulk parenteral product leaving the facility is rarely sterile (it is terminally sterilized or sterile filtered in the pharmaceutical plant), often Sterilize-In-Place systems are needed for certain equipment.

In a BPC manufacturing facility, costs associated with open processes are driven not only by GMPs, but also by environmental, insurance, worker protection, and energy issues. Closing of bulk manufacturing processes solves many of those problems.

IV. ORAL SOLID DOSAGE (OSD) FACILITIES

The *Oral Solid Dosage (OSD) Baseline Guide* (2) was created with many of the same team that participated in the *BPC Baseline Guide*. A broader cross section of the pharmaceutical industry reviewed the guide, with considerably more discussion as the industry began to realize the potential benefits of defining baseline facility expectations for meeting GMPs. This guide was published in 1998.

The *OSD Baseline Guide* covers the manufacture of dry products meant to be taken orally by the patient. Such products are formulated and finished in the form of tablets, coated tablets, powders, and capsules. Process steps include dispensing, blending, granulation, drying, milling, compression, and filling.

Since product sterility is not an issue, the emphasis is to protect the OSD product from undue contamination from other products and from excessive bio-contamination. Since OSD facilities often process multiple products, product cross-contamination is a major issue. It would be wishful thinking to expect a manufacturer to know how much of a Product A can be tolerated in Product B, C, D, or E, so the emphasis on the facility design is to prevent undue contamination. Containment by negative pressure creating proper airflow direction is suggested for open processes. But as before (in the BPC plant), if a process is closed and can be kept closed, such cross-contamination concerns diminish.

Because dry product can be more difficult to clean, and because every step of the process is a critical step, residues left in equipment place a heavier burden on the cleaning regimen. If equipment can not be cleaned in place (i.e., "closed"), it may either be cleaned in a dedicated negative pressure facility, or left in place in the processing room and cleaned in an open manner. If this cleaning process involves dislodging dust to the room environment, negative room pressure is required and often some sort of dust capture is necessary. If the product is a threat to workers, PPE will be required.

Again, if a process can be closed or isolated, as with validated glove box processes, and cleaned in place, GMP concerns become less difficult to satisfy.

V. STERILE MANUFACTURING FACILITIES

The *Sterile Manufacturing Facilities Guide* (3) required the largest amount of writing effort to date. A core team from several European manufacturing companies who are subject to FDA inspection created the guide over a time frame of almost three years. Because of the nature of the products covered, and because FDA is currently rewriting the *Aseptic Processing Guideline*, extensive review by industry and the agency was required. The guide, published in 1999, includes appendices on European GMPs and further "how-to" information on HVAC design.

In sterile product manufacture, the product may be processed aseptically, requiring the process to be performed in increasingly clean environments. In addition, equipment in contact with the product and water used in the product must meet rigorous particulate, bioburden, and endotoxin requirements. Open filling of the sterile product must be performed in a Class 100 environment inside a Class 10,000 clean room. Operators must be carefully and heavily gowned, and must follow controlled procedures to prevent contamination of the exposed product, equipment, containers, and closures. Not only must the filling area be tightly monitored, but the room environment must be as well.

Other sterile products are "terminally sterilized," meaning that product and contact surface controls can be somewhat relaxed, since the product will be sterilized after filling and stoppering, usually by heat. This would seem to be a desirable situation, but most parenteral products, especially biotech products, can not tolerate heat in their final form.

In addition to the chapters mentioned previously, the *Sterile Manufacturing Facilities Guide* includes a chapter discussing current approaches to "barrier isolator" technology. Although it addresses a moving target, it provides a number of considerations for the process designer and was, at the time of its writing, aligned with FDA expectations for sterile barrier isolator operations. Of foremost benefit to the facility owner is a relaxing of the filling room environment to Class 100,000. Although not a "shirtsleeve" environment, this room will require less extensive gowning and monitoring than a traditional Class 10,000 filling room.

A barrier isolator separates the operator from the product, essentially eliminating contamination of the product by the single greatest source of product contamination, the operator (note that an overzealous operator can unintentionally contaminate a barrier, so controlled operating procedures are still necessary). If the product requires low humidity (such as a sterile powder) the humidity control system serves the inside of the barrier only, leaving the operator in a more comfortable environment outside the barrier. The cost of the building and operating facility can be greatly reduced, but the cost of the process inside the barrier increases. For example, sterilization of the barrier interior will require an in-place system, with rigorous validation. As more manufacturers develop barrier isolator processes, the cost of barriers should continue to come down.

Of particular note to the hygienist is the filling of highly potent sterile products in a barrier isolator. Often the barrier is not a tightly closed glove box, but it has openings (''mouse holes'') for vials to enter and for stoppered vials to leave. Other items, such as stoppers and tools, can be passed into the barrier through transfer ports that act as airlocks to prevent a direct path between the barrier environment and the room outside. However, since the barrier is under 0.05 in., (~12 Pa) or more pressure, any airborne product inside the barrier will be blown into the operator's work environment through the exit mouse hole (see Figure 1). If the product is a potent sterile powder, this can be a major concern. Approaches to solving this problem have included spot exhaust capture at the openings (being careful not to disrupt the Class 100 environment over the stoppered vials on their way to the capping machine). Another approach is double-walled barriers such that the interstitial wall space pulls air from both the barrier and from the exit area environment.

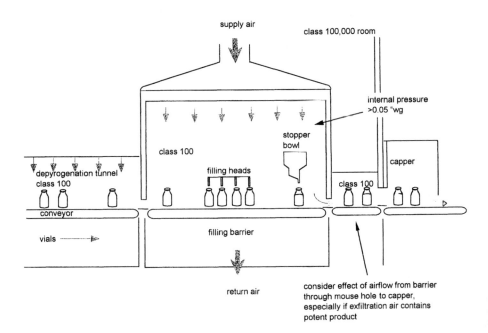

Figure 1 Aseptic filling barrier.

Many factors, such as sterilization-in-place, dehumidification, glove integrity, airflow pattern testing, drainability after washing, product and component access, and barrier leak testing present special challenges to the process designer. Discussion of these considerations is provided in more detail in the *ISPE Sterile Manufacturing Facilities Guide* and in trade journal articles by barrier experts.

Interest in barriers as an alternative to traditional clean room operations continues to increase. The International Standards Organization (ISO) has convened a working group to develop an ISO standard for "Enhanced Clean Devices." This standard will cover the cleanliness aspect of barriers, but may also touch on worker-protection issues.

Although barriers present an attractive alternative to open sterile filling, they may not prove to be the most cost-effective approach for a particular process. However, when sterile potent compounds, especially powders, are involved, a barrier isolator may provide a practical alternative. Because the GMPs and the technology for barrier operations are constantly evolving, pharmaceutical manufacturers considering their first barrier isolator installation should consult with as many resources as possible, and especially discuss their plans with regulatory agencies before investing. There have been a number of very informative articles written by experts, among them FDA Compliance Officers, some of which are listed in the reference section of this book (4).

Facilities used for drug development, where the product is not intended for human use, may have relaxed GMP requirements. Sterility requirements usually are not as strict, since the intent of a development facility is to learn about the product: how to make it, how it affects people, how it acts in the workplace, etc. However, should the product developed in a pilot plant be used for human clinical trials or for product launch, GMPs should align with full production requirements. Since a new drug's potency may not be well known, worker protection is a high-level concern. Often the development of new, relatively unknown compounds requires attention to worker protection as well as some level of product protection to prevent contamination.

VI. FUTURE GUIDES

The *Baseline Guides* written to date have concentrated on small molecule pharmaceutical products (i.e., drugs) that fall under the Center for Drug Evaluation and Research (CDER) branch of the FDA. Guides currently in progress will cover Water and Steam Systems, and Commissioning and Qualification, and will apply to all FDA-regulated pharmaceutical manufacture. The *Baseline Guide for Biotech Facilities* will point out differences between drug and biotech operations and will have considerable input from the FDA Center for Biologics Evaluation and Research (CBER) and Team Biologics inspectors. Biotech products create a new set of problems, in that the large molecules are especially sensitive to biocontamination. Many of them are also very powerful and create a risk for the process operator. These compounds may be harmful if released to the building environment or to the outdoors. As stated previously for drug manufacturing, containment of the process and control at the source could very well be the answer to a number of problems.

For more information on the *ISPE Baseline Guides,* contact ISPE, or visit their Web page at www.ispe.org

REFERENCES

1. Bulk Pharmaceutical Chemicals, Baseline Guide for New Facilities. ISPE, June 1996.
2. Oral Solid Dosage Forms, Baseline Guide for New and Renovated Facilities. ISPE, February 1998.
3. Sterile Manufacturing Facilities. In Baseline Guide for New and Renovated Facilities. ISPE, January 1999.
4. Richard L Friedman, FDA Office of Compliance. Design of Barrier Isolators for Aseptic Processing. Pharmaceutical Engineering, March/April 1998.

3
Industrial Hygiene Aspects of Pharmaceutical Manufacturing

Donna S. Heidel
Johnson and Johnson, New Brunswick, New Jersey

I. INDUSTRIAL HYGIENE DEFINITION

The science of industrial hygiene has been described as "the anticipation, recognition, evaluation and control of those environmental factors or stresses, arising in or from the workplace, which may cause sickness, impaired health and well-being or significant discomfort and inefficiency among workers or among the citizens of the community" (1). Pharmaceutical compounds are considered to be a chemical hazard. However, unlike most chemical hazards, pharmaceutical compounds are designed to have a biological effect at very low dosages. Although this biological effect is considered to be beneficial to the patient, any biological effect in the worker is undesirable. In addition, although pharmaceutical compounds undergo rigorous safety testing, toxic effects, such as reproductive toxicity, mutagenicity, and allergic reactions, have been reported.

II. ROUTES OF ENTRY

A. Inhalation of Pharmaceutical Compounds

The primary route of exposure to pharmaceutical compounds is through inhalation. Generally, pharmaceutical compounds can be inhaled as either dusts or mists. Dusts, or finely divided powders, are frequently generated and released as part of both bulk pharmaceutical chemical synthesis and formulation. During synthesis, dusts are generated and released during centrifuging, drying, milling and weighing activities. During formulation, sampling, weighing, dispensing,

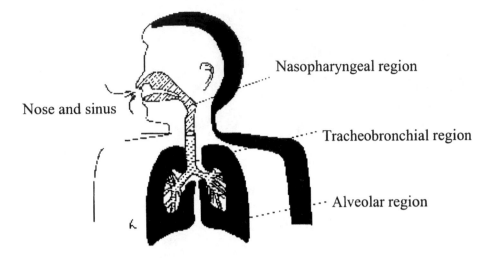

Figure 1   The human respiratory tract.

charging, granulation, blending, milling, or sieving, tablet compression and even tablet coating can release significant amounts of finely divided powders.

Drug-containing mists can be released into the employee's breathing zone during sonicating and vial filling. Opening formulation equipment during and after drying cycles can cause the employee to inhale drug-laden steam. Finally, cleaning equipment and rooms with high-pressure hoses can cause the inhalation of drug-containing mists.

Deposition in the respiratory tract depends on particle size. Many nonpharmaceutical dusts are of concern when particles are less than approximately 10 μm in diameter since these particles penetrate deeply into the alveoli and either enter the systemic circulation or damage the alveolar membrane. Pharmaceutical compounds, however, can be absorbed systemically even with particle sizes of up to 100 μm in diameter. Like nonpharmaceutical particles, pharmaceutical particles of less than 10 μm in diameter can enter the systemic circulation through the alveolar membrane (see Figure 1).

Pharmaceutical compound particles ranging in size from 10 to 50 μm that are too large to enter the alveolar region of the lung are usually deposited in the bronchi in the thoracic region. The cilia in the bronchi clear the lungs of the compound by rapidly moving back and forth. This "mucociliary escalator" raises the particles to the throat where they are removed by coughing and swallowing. If swallowed, the compound can then enter the body through the oral route.

Finally, the nose and sinuses of the upper respiratory tract usually filter out compounds that are between 50 and 100 μm in diameter. However, compounds that are soluble in the moist environment of the upper respiratory tract can enter the systemic circulation through absorption across the mucous membranes. As a result, size-selective sampling for pharmaceutical compounds is generally not recommended. Most pharmaceutical compounds are sampled by collecting all particles less than 100 μm (3).

B. Dermal Absorption of Pharmaceutical Compounds

Pharmaceutical compounds may function as primary irritants, can cause allergies, or can be absorbed systemically across the skin. The skin as a route of exposure is generally not considered when an occupational exposure limit (OEL) is established. However, skin absorption can be a significant contributor to an employee's overall exposure to a pharmaceutical compound and should not be overlooked. Pharmaceutical compounds can be absorbed systemically via the following mechanisms: transdermal or across the skin, percutaneous through injection by needles or other sharps, and across the mucous membranes of the eyes and upper respiratory tract.

Many pharmaceutical compounds can be absorbed directly via the skin, even as dry powders. This transdermal absorption can be intensified by drug delivery systems designed to be worn as skin "patches." Obviously, these formulations can also be inadvertently absorbed through the skin of employees during processing.

Pharmaceutical compounds can be absorbed systemically through contact with contaminated needles, broken glass, or lacerations from contaminated equipment. Compounds that can affect skin integrity or can be absorbed through the skin should be designated through the use of a "skin" notation. Wipe, or "swab," testing may be useful to evaluate the potential for skin absorption in the work setting.

C. Ingestion of Pharmaceutical Compounds

We generally do not consider the ingestion route in an occupational setting. However, ingestion of pharmaceutical compounds can occur in two ways. First, hand-to-mouth contact from eating, drinking, or applying cosmetics with contaminated hands can result in ingestion of the drug substance. In addition, placing contaminated articles, such as pens and pencils, into the mouth can cause ingestion of a drug.

As we discussed earlier, drugs can also be ingested through the "cough and swallow'' reflex from bronchial deposition. The drug can then be absorbed systemically through the GI tract.

III. OCCUPATIONAL EXPOSURE LIMITS

The pharmaceutical industry has a long tradition of establishing internal occupational exposure limits for pharmaceutical compounds. Although not regulated by the Occupational Safety and Health Administration (OSHA) domestically or the equivalent internationally, pharmaceutical drugs are by their very nature designed to have significant biological effect at relatively low dosages by efficiently entering body systems and binding to certain receptors. While these biological effects are intended and very beneficial in the patient, they are very undesirable, and may even be harmful, in the employee synthesizing the drug substance or formulating the drug into dosage forms. The industrial hygienist must also consider that employees working with pharmaceutical compounds may take other drugs therapeutically. Both the industrial hygienist and the occupational health professional must understand that drug exposures in the workplace can add to, be synergistic with, or even potentiate drugs the employee takes therapeutically. In addition, taken together, some drugs can cause toxic interactions, such as cardiac arrhythmias.

Some drug substances may also exert a toxic effect, such as a reproductive toxic effect. They may also be mutagenic or carcinogenic. Reproductive toxicants are compounds that may affect libido or fertility, cause abortions or birth defects, or result in developmental effects as the child grows and develops physically or mentally. Mutagenic drugs can damage the DNA and may result in inheritable genetic effects. Carcinogenic drugs can result in the development of cancers.

Occupational exposure limits can be defined as the time-weighted average concentration of a drug, measured in the employee's breathing zone, that is considered to be safe for the majority of healthy workers, for an 8-hour shift and 40-hour work week. OELs are generally calculated as 8-hour, time-weighted average concentrations, although some pharmaceutical companies calculate OELs to accommodate 10-hour or 12-hour work shifts.

Establishing an OEL requires a team of scientists representing the following disciplines: toxicology, pharmacology, organic chemistry, clinical physicians, occupational physicians, occupational toxicology, and industrial hygiene. It is also important to include pharmaceutical process engineers and operations managers since they will inherit, and must ultimately contain to, the OEL.

The OEL team considers the following data, derived from animal toxicology and human clinical studies, to establish the OEL: therapeutic dosages for human use; animal toxicology studies; and human pharmacology, pharmacokinetic, and pharmacodynamic studies.

The OEL can be calculated using the following formula:

$$\text{8-hour TWA} = \frac{\text{NOEL (mg/kg/day)} \times \text{BW (kg)}}{\text{V (m}^3\text{/day)} \times \text{S (days)} \times \text{SF} \times \alpha} \tag{1}$$

where NOEL = no observable effect level (LOEL, or Lowest Observable Effect Limit, can be substituted if NOEL unavailable); BW = Body Weight for a typical employee, generally between 50 and 70 kg; V = 10 m^3 (2). This is the amount of air breathed during moderate work over an 8-hour day. This factor can be adjusted to achieve a 10-hour or 12-hour work shift; S = time, in days, to achieve plasma steady state; SF = safety factors, based on toxicology assessment; α = fraction of compound absorbed via inhaled route. Note, there are other acceptable methods or formulas also used in the pharmaceutical industry as well.

As mentioned previously, this 8-hour, time-weighed average concentration can be expanded to a 10-hour or 12-hour time-weighted average, depending on the work shift, as long as the ''safe dose'' is not exceeded.

OELs can also include a short-term exposure limit (STEL) for compounds that may cause acute effects from short-term exposures. Generally, the STEL is approximately three times the OEL and is measured as a 15-minute time-weighted average.

Following the establishment of an OEL, a document or monograph should be written and approved. Creating a monograph that defines the factors considered when developing an OEL is recommended to serve as an approval and communication document as well as a record of factors included if changes to the OEL are considered in the future.

IV. INDUSTRIAL HYGIENE SAMPLING

Following the establishment of an OEL, or concurrent with the process, an industrial hygiene sampling and analytical method is developed. Industrial hygiene sampling is a process that quantifies the breathing zone concentration of a drug-active ingredient during a work activity. Before the industrial hygienist can begin to assess employee exposure, a laboratory experienced in the analysis of industrial hygiene samples of drug actives must develop an analytical method. Generally, development of this method should include the following factors: developments of calibration curves, stability of the drug on filters, target limit of quantification (at least 0.1 × the OEL), desorption efficiency, sampling efficiency, and analytical precision and accuracy.

Following development of the analytical method, industrial hygiene sampling can begin. Samples are obtained by drawing a known volume of air across a filter using a calibrated sampling pump. The filter is contained within a sampling cassette and is placed in the employee's breathing zone, generally on the lapel of the work uniform. The employee wears the sampling train for the entire work activity or work shift, depending on the intent of the survey. Short-term sampling during a single work activity assists the industrial hygienist and the engineer in "diagnosing" specific unit operations that may require additional levels of containment.

Sampling for the entire work shift evaluates the employee's total exposure to the drug during the shift. Whether the strategy includes diagnostic or full-shift sampling, enough samples must be gathered for the survey to have statistical relevance.

At the conclusion of the sampling survey, the samples are sent to a laboratory with experience in analyzing for drug actives. The laboratory results are compared against the OEL. If the results are greater than 0.5 of the OEL, measures to reduce the employee's exposure are taken. Throughout the world, reducing employees' exposure to a chemical agent is accomplished according to the following priority: elimination of the material, substitution of the material, engineering controls, administrative controls, and personal protective equipment. Generally, elimination and substitution are not possible in pharmaceutical manufacturing. Using engineering controls, such as local exhaust ventilation, high-containment valves for materials transfer, modification of process equipment, or glove box isolators, becomes our preferred method of control.

Administrative controls, such as job rotation or work practice modifications, are valuable tools in reducing employee exposure. However, they are very difficult to administrate since they require very thorough understanding of each employee's work practices and the resultant airborne concentrations, tight control of employee work activities, and continuing retraining on work practices required to keep exposures below the OEL.

The least desirable method for managing employee exposure to pharmaceutical compounds is through the use of personal protective equipment, such as respirators. Respirators and protective clothing do not alter the airborne concentrations. They simply filter the drug from the employee's nose and mouth. Personal protective equipment is undesirable since it is associated with a relatively high rate of failure. In addition, using respirators requires that employees are continually retrained, and medically examined and approved to wear respirators. Employees generally do not like to wear respirators since they are hot, increase the work of breathing and reduce their field of vision. Also, using respirators means the engineer or industrial hygienist must establish both a process and area for routine inspection, repair, and decontamination of the respirators.

V. LIMITATIONS TO THE OEL/SAMPLE/RETROFIT APPROACH TO MANAGING EXPOSURES

This traditional industrial hygiene approach for managing employee exposures in the pharmaceutical industry has significant limitations, especially in today's pharmaceutical R&D arena. First, since an OEL is set after human clinical trials, NDA filing, and oftentimes chemical and pharmaceutical process design, it is established too late in the pharmaceutical development process to affect process containment. Therefore, it relies on the industrial hygienist to find exposure problems and correct them, rather than anticipating and preventing them. Second, until retrofits can be designed and implemented, exposures are managed through the use of personal protective equipment. Since personal protective equipment has a significant rate of failure, employees are at risk for health effects from exposure to the pharmaceutical compounds. Third, simply providing a numerical OEL to research and development scientists, process engineers, operations management, and facilities engineering is only a beginning, not an end unto itself. The OEL does not assist the chemical or pharmaceutical engineer in designing processes to contain to below the OEL. It only gives a target to shoot for. Lastly, until the industrial hygienist can fully characterize the process, it may result in significant exposures to the employees.

VI. PERFORMANCE-BASED OCCUPATIONAL EXPOSURE LIMITS (PB-OELs)

Using a performance-based approach for managing employee exposures corrects the shortcomings of the traditional industrial hygiene process. Performance-based Occupational Exposure Limits, or PB-OELs, offers a systematic method of assigning compounds to one of up to five categories based on potency, pharmacological effect and toxicological effects. Established early in preclinical development, PB-OELs result in an order(s) of magnitude estimation of the OEL. An OEL is still needed for each compound and is established later in the development of the drug once human clinical data is obtained. Each PB-OEL category prescribes the appropriate facility and process containment strategies effective in reaching OELs contained within the category.

A. PB-OEL Process and Benefits

The goal of PB-OELs is to categorize compounds within the first six months of development. As with OELs, compounds are categorized into the appropriate containment category by a multidisciplinary team of R&D scientists. Compound categorization is based on potency, expected therapeutic dose, pharmacology, genotoxicity, acute toxicology, benchmarking of similar compounds, and therapeutic class.

The business need for PB-OELs is significant. One of the most important reasons to adopt a performance-based approach for managing employee exposure is that pharmaceutical compounds are becoming increasingly potent. Obviously, as potency increases, OELs decrease. The traditional industrial hygiene method of sampling employees' breathing air is less viable for highly potent compounds due to constraints from the detection limits of standard HPLC analytical methods.

Second, facility retrofits are difficult and costly. Time is money, and waiting during facility or process start-up to determine that the containment is not adequate to achieve the OEL results in very expensive process revisions that often require filing with the appropriate regulatory authority. As a result, significant retrofits are often out of the question.

Third, facilities and processes are not overdesigned. More extensive containment technology is used only when necessary.

Fourth, PB-OELs provide a tool for operations management to consistently select a manufacturing facility capable of meeting the OEL.

Finally, the most important business reason to adopt a performance-based approach is to prevent occupational illness. Designing the appropriate level of containment into the process and facility will significantly reduce the possibility of occupational illness, especially if the containment strategy does not rely on personal protective equipment.

The outcome of a PB-OEL approach is containment that is based on the use of engineering controls. This is important not only in the prevention of occupational illnesses, but also in reducing the cross-contamination risk in manufacturing facilities. Using process containment, rather than air changes and filtration systems as the primary means of controlling exposure risks also reduces energy and facility cleaning costs.

B. Containment Strategies for Each Containment Category

In order for this approach to be most effective, the containment categories should be based on the capabilities of the engineering containment controls. One company's approach, briefly summarized in the following table, uses four containment categories increasing in potency from low to extreme, based on the daily normal dose the patient will typically receive.

Category	Potency	Design	OEL range
1	Low >100 mg/day	Conventional open equipment; incidental contact with compound	>100 ug/m^3
2	Moderate 10–100 mg/day	Gasketed, flanged equipment; laminar flow/directional laminar flow; enclosed transfers	20–100 ug/m^3
3A	High 0.01–10 mg/day	Transfers using high-containment valves (e.g., split butterfly valves); containment for every disconnect	20–5.0 ug/m^3
3B		Barrier technology (negative pressure)	<5.0 ug/m^3
4	Extreme <0.01 mg/day	Remote operation; fully automated; no human contact	<0.01 ug/m^3

Each category specifies appropriate exposure control to meet OELs in that range. In this company's approach, compounds that have not yet been categorized are handled using Category 3 criteria. The recommended technologies in each containment category are based on the known capability of the equipment.

A comprehensive PB-OEL program should include the following elements: facility and mechanical systems, process equipment and containment, equipment cleaning and maintenance, employee work practices, laboratory containment and practices, industrial hygiene programs such as sampling and the use of personal protective equipment as a redundant control, and health surveillance programs.

C. Specifying and Testing Containment

Achieving the OEL for new process equipment requires that equipment specification include the appropriate containment requirements. The design team should review the containment requirements for the process and establish a design exposure limit. The design exposure limit is the breathing zone concentration of the drug during the unit operation, such as charging a process vessel or operating a sieve. All equipment should contain this containment specification.

Rather than waiting until process validation to conduct industrial hygiene sampling on new equipment, sampling should occur during factory acceptance testing and operational qualification as well. An appropriate surrogate should be selected for the testing. An analytical industrial hygiene method should be developed with a sufficiently low detection limit to allow for short-term sampling. Sampling results should be compared against the design exposure limit. Testing earlier than process validation will allow for minor equipment modifications that were not anticipated during equipment specification. Traditional industrial hygiene sampling for the compound should be a part of commissioning. Depending on the results of the survey, ongoing industrial hygiene sampling may be required.

REFERENCES

1. JB Olishifski. Fundamentals of Industrial Hygiene. Chicago: National Safety Council, 1971.
2. EV Sargent, GD Kirk. Establishing Airborne Exposure Control Limits in the Pharmaceutical Industry. American Industrial Hygiene Association 49:309–303, 1988.
3. JH Vincent. Particle Size-Selective Aerosol Sampling in the Workplace. American Industrial Hygiene Association, 1996.

4

Effect of Individual Particle Characteristics on Airborne Emissions

Chetan P. Pujara
Abbott Laboratories, North Chicago, Illinois

Dane O. Kildsig
Purdue University, West Lafayette, Indiana

EDITOR'S NOTE

Interest has increased recently in studying the actual dustiness of pharmaceutical materials. In terms of containing potent compounds, the dustiness level that a compound inherently generates by virtue of its own physical properties is of immediate interest since that will directly drive the challenge to the containment system that is put in place to contain it. The logical end to this area of inquiry is to be able to predict how dusty a hypothetical powder will be, based on its planned physical characteristics, and then to select or synthesize dry pharmaceutical ingredients to meet these lower-dustiness physical profiles. In so doing, a large portion of the traditional containment challenges would be mitigated before the compound ever made its way to the production floor. This is exciting research with some interesting implications.

I. INTRODUCTION

Powders and granulated solids are used throughout the pharmaceutical industry. The handling of these materials generates airborne dust that may affect worker health and safety, cause a nuisance and/or result in product loss. This is especially true when the dust is an active chemical ingredient. Dust is defined as any particulate material finer than 75 μm (1). *Dustiness* (or *dustability*) is defined as the propensity of a material to emit dust during handling operations and may be considered to be analogous to vapor pressure on the molecular scale. The process by which dust is produced is referred to as pulvation and is analogous to evaporation on the molecular scale (2). Containment technology in the pharmaceutical industry would benefit from a systematic study of powder dustiness and the powder factors governing the aerosolization and transport of the airborne particles. The methods of preventing dustiness are of increasing importance in handling of powders due to the growing emphasis on health and safety, and also on loss prevention.

Dustiness studies can be used in many ways (3, 4):

1. Determination of worker exposure: When potent compounds are involved, it is necessary to eliminate the exposure of these compounds to workers. The drugs being manufactured by the pharmaceutical industry are becoming increasingly potent. So the problem needs to be addressed again with special attention to these newer more potent compounds.

2. Quantification and estimation of product loss: Product loss due to dust emission can result in significant increases in cost of manufacturing and handling of expensive drugs in the industry.

3. Assessment of the relative dustiness of a material and determination of the need for dust control: Dust may be classified as lung-depositing dust, toxic dust, primary-irritant dust, sensitizing dust, or nuisance dust with regards to worker's health. Particles that are less than 10 μm in diameter are usually considered as lung-depositing dust. Toxic dust is a systemic poison that enters the circulation through the oral, pulmonary, transdermal, or other routes. Primary-irritant dust is limited usually to the eyes, nose, and throat. Prolonged exposure to irritants can cause alterations in respiratory function. As a result of inhalation, skin contact, or ingestion, a worker may become sensitized to substances. Finally, nuisance dust is discomforting to the worker and is often associated with increased colds and bronchitis (5).

4. Assessment of the effectiveness of dust suppression techniques.

5. Quality control by production of material/product with decreased dustiness.

II. METHODS TO DETERMINE POWDER DUSTINESS

Many devices have been used to determine the dustiness of bulk powders. Standardization of both the method and the index is essential for dustiness index values to be meaningful. (See the Appendix at the end of this chapter for discussion and definition of the dustiness index.) Unfortunately, there are numerous dustiness-estimation methods being used today. There have been attempts to standardize dustiness tests and indices but these attempts have failed considerably. The ubiquitous nature of the problem and the absence of a standardized test has led to the construction of many indigenous devices to determine dustiness. Thus, instruments have been classified into three broad categories based on their method of dispersing the powder: gravity dispersion (drop tests), mechanical dispersion, and gas dispersion (fluidized beds).

A. Gravity Dispersion

This class of method allows a mass of product to fall into an enclosed space, usually a box-shaped chamber or to be tipped from a container within the chamber. Further classification may be obtained according to the method used to evaluate the resultant dust.

1. Mass Determination

Determine the mass of the amount of dust generated using a membrane or glass-fiber filter to collect the dust. The dust is usually collected on the filter using a suction pump operating at a fixed flow rate (Active Sampling). An alternate arrangement is to sample passively by placing the filter at various locations in the chamber and allowing the dust to settle onto the filter. The ASTM standard instrument for determining coal dustiness (6) uses a passive sampling method. In the case of active sampling, a cascade impactor may be attached to the instrument to simultaneously obtain particle size distributions and the mass of the dust generated.

2. Light Obscuration

The quantity of dust is assessed by the use of a beam of light. Mikula and Parsons (7) used a helium-neon laser as a light source with a single photodiode as the detector. The intensity of the light was recorded continuously as the dust settled after a known quantity of coal was dropped in a chamber.

   There are several reports on development and use of instruments that employ gravity-dispersion methods to determine powder dustiness (5, 8, 9, 10). The more commonly reported instruments are the Perra pulvimeter (11), Vertical flow dust chamber (12), Modified Perra pulvimeter (13), Laboratory dust disperser (14), and Midwest Research Institute dustiness tester (15).

B. Mechanical Dispersion

Instruments that employ this method of dispersion generally contain a drum with baffles in which the powder to be tested is placed (4, 16, 17). The drum is rotated to disperse the powder and the airborne particles generated in the drum are carried away by pulling air through the drum. The aerosol is then analyzed gravimetrically or by light-scattering devices as described under gravity-dispersion methods. The Warren Spring Laboratory dustiness tester (18) was developed in 1981 and the Heubach Dustmeter was developed in 1984 (16). These are the two major rolling-drum testers used today.

Castor and Gray (19) used a very unique instrument that employs a spring-loaded striker that hits a glass cuvette (placed on an anvil) from the bottom. The cuvette contains the powder sample and when hit, the powder generates dust and a photon correlation spectrometer measures the amount of light scattered by the dust. This method has the disadvantage of powder particles adhering to the walls of the cuvette, which leads to false measurements of dustiness indices.

C. Gas Dispersion

In this methodology, dust is liberated by the action of a gas passing through the sample. Fluidized bed testers have been used in the past but currently are not being used much in the estimation of dustiness. There are few reports in the literature that have used gas dispersion to determine powder dustiness (7, 20, 21).

III. DUSTINESS TESTING: WORKING TOWARD A SINGLE METHOD

The British Occupational Hygiene Society established a working group in 1981 to develop a procedure to measure dustiness, to establish a dustiness index scale, and to correlate the measured dustiness with actual worker exposure (18). The group evaluated 18 measurement devices but no dustiness index scale or standard methodology was established. However, the working group did decide to eliminate gas-dispersion methods as impracticable in relation to a standardized method for general application. Also, the group decided to focus on gravity (drop tests) and mechanical (rolling-drum tests) dispersion methods to provide a standard method for dustiness measurement. Chung and Burdett (22) also proposed the standardization of dustiness tests and indices after a review of several dustiness measurement techniques. A standard test method (rolling drum) based on their index was suggested for future studies.

Higman (3) evaluated different dustiness testers and the summarized results indicated that a rolling-drum tester is more versatile than a drop test or a fluidized bed device. In the first two categories, the Heubach and Midwest Research Institute (MRI) dustiness testers are the most commonly used for determining powder dustiness.

The Heubach Dustmeter, a rolling-drum tester, is a commercially available instrument in Europe (16). A bench-scale impact-type chamber called the MRI Dustmeter (gravity dispersion) was developed to measure the dustiness of finely divided materials (15). Extensive research on powder dustiness has been done with these two dustiness testers.

A significant correlation was observed between dustiness test results (using MRI and Heubach Dustmeters) and dust exposures in a packaging room for a powdered acrylic-resin production line (23). The correlation between dustiness test results and worker dust exposure at two bag-filling and two bag-dumping operations in lead chromate, powdered acrylic resin, and paint plants was also evaluated (24, 25). The results from two sites (A and B) suggested significant correlation between dustiness test results and dust exposures in the plant operations. At the two other sites (C and D) there was a poor relationship between worker dust exposure and dustiness test results. The exposure at sites C and D was very low when compard with sites A an B because less dusty materials were encountered at sites C and D.

Instrument factors that affect dust generation in silicon carbide and aluminum oxide powders were examined using the MRI and Heubach Dustmeters (26). The authors recommended the use of only one dustiness tester for future research and suggest using the Heubach tester since it requires a sample only one-eighth of the amount required for the MRI tester.

A bench-top apparatus was built to examine factors that affect dust generation (10). This tester employs material impact following a free fall as the mechanism of dust generation (gravity dispersion). The dusty air is pulled at 10 L/min. through an elutriation column which is connected with a five-stage impactor for the collection of dust particles according to their aerodynamic size. The authors conclude that the new tester agrees with previous research (Heubach and MRI tests) in most cases, and that the results are reproducible.

There are many different ways of measuring dustiness, as evident in this review, depending on the material under test and the type of process being simulated. Extensive literature studies indicated the Heubach and the MRI testers are the most popular dustiness-testing devices. The Heubach dustiness tester is a commercially available instrument and has been used extensively to determine the dustiness of powders. This tester has proved to be a reproducible and reliable instrument for routine dustiness tests. It also requires a much smaller sample size than the MRI tester. Substantial research has been done with this device, as discussed in this section, to study factors that affect dustiness. The Council of the European Communities (27, 28) has adopted the Heubach device as a standard to determine the dustiness of feed premixes used in the agricultural industry. These published reports suggest that the Heubach dustmeter may be the ideal instrument for dustiness testing of pharmaceutical bulk powders.

IV. RELATIONSHIP BETWEEN DUSTINESS AND POWDER PHYSICAL CHARACTERISTICS

Powder dustiness is a composite property dependent upon many variables. A majority of the earlier work revolved around nonpharmaceutical materials such as limestone, coal, and alumina. Recently there has been increased interest generated in studying dustiness of pharmaceutical materials. In the following sections, a review of the effects of powder parameters on dustiness of both pharmaceutical and nonpharmaceutical materials is presented. Specifically, the effect of particle size and size distribution, particle shape, powder mass, bulk and true density, flowability, cohesion, and moisture content have been discussed.

A. Powder Mass and Bulk Density

The mass of powder placed in a dustiness tester does have an effect on the percent of dust generated from the sample. This is true for both the drop and rolling-drum testers. Bulk density of a sample needs to be considered primarily for the drop tests only, since dust is generated by allowing the entire powder mass to fall only one time. In a rolling drum tester changing the bulk density of a sample does not have a great impact because the powder is repeatedly picked up by baffles and dropped for a set amount of time causing aeration of the powder.

Increasing the sample mass reduced the dustiness of alumina in a laminar flow apparatus (29) that can be used to release a sample as a mass or as a stream from a fixed height. This could probably be due to the fact that as sample size decreases, the forces tending to bind fine particles together are more easily overcome by the shearing forces created as the sample drops through the air. Also, using a gravimetric apparatus they showed that dustiness decreased with an increase in mass of chalk powder although they did observe an initial increase in dustiness when sample mass was increased.

Dustiness of alumina was studied using a modified perra pulvimeter (13, 30) The results indicated that dustiness increased with increase in bulk density of the powder.

Castor and Gray (19) studied the dustiness of lead chromate in a domestic dustiness tester based on a light-scattering measurement process. Results indicated that the scattering maximum (dust) increased linearly with an increase in the weight of the lead chromate powder. This method of dustiness testing showed some problems of powder adhering to the sides of the dust-generating chamber, which compromised the test results.

The factors that affect the Heubach and MRI dustiness tests were studied using powdered limestone (24). The Heubach dustiness index increased with increasing mass up to ~80g of sample weight and then the dustiness index decreased with increase in sample weight. However, the differences between dustiness test results at 10 and 20 g were not significantly different from each other. In the MRI test, increasing the bulk density from 0.73 g/cc to 1.07 g/cc significantly decreased the number of particles between 5 and 17 µm collected on the filter but appeared to increase the number of particles in the 1 to 5 µm range. This caused the MRI index to decrease with increasing bulk density.

Seven pharmaceutical powders were used to study the effect of sample mass on powder dustiness index in the Heubach Dustmeter (31). The dustiness index of these powders at 10 L/min. airflow rate increased initially with powder mass and decreased with further increases in powder mass. The dustiness index of powders decreases with increasing sample weight probably because the volume of the powder falling to the bottom of the dust generator increases with increasing weight, which would decrease with air entrained, resulting in lower dust generation. On the contrary, the amount of material impacting the bottom of the drum increases with weight since the drum baffle picks up more powder, which could result in an increase in the amount of powder rising from the bottom of the dust generator. However, not all the particles rising up will be picked up by the airstream because the particles will collide with each other due to increased concentration of particles at higher sample mass. Therefore, both the instrument and the process of dust generation itself affect the amount of dust collected on the filter as the weight of the powder placed in the dust generator is increased. A 10 g sample was found to be optimal for dustiness tests in the Heubach Dustmeter.

B. Moisture Content

Most studies have shown that increasing the moisture content of a bulk powder decreases its dustiness. This is a well-understood concept since moisture increases the cohesion of powders and increases the weight of the particle. Consequently, powder dustiness will generally decrease with an increase in moisture content of the powder.

Using a fluidized bed device, it was shown that increasing the moisture content of a material from 10% to 15% reduces its dustiness by a factor of 7 (2, 20).

In a modified Perra pulvimeter, it was found that the dustiness index of alumina increased from 2.5 to 3.5 (g dust/kg alumina) when the moisture content was increased from 0% to 1% but a further increase in moisture content led to a steady decrease in the dustiness index (13).

An MRI tester was used to relate the dustiness of powders to their physical characteristics (15). Bulk density, particle size distribution, angle of repose, and moisture content of the powder samples were determined along with the powder dustiness. Dustiness of all the powders decreased with increase in moisture content except for powdered NaCl and Talc. For NaCl and Talc the humidified sample dustiness was two to four times the dried sample dustiness. Stepwise multiple-linear regression was used to obtain the following equation,

$$L = 16.6(M)^{-0.75}(S_g)^{3.9}(D)^{-1.2}(M_g)^{-0.45} \tag{1}$$

where, L = fractional mass loss (mg/Kg), M = moisture content (%), D = bulk density (g/cc), M_g = mass median diameter of the particle size distribution, and S_g = standard geometric deviation of the particle size distribution.

Mikula and Parsons (7) used a funnel test apparatus and a light transmittance test to characterize coal dustiness instead of the standard method suggested by ASTM D547 (6). In the funnel test, 50 g of coal was placed in a fritted glass funnel and air was passed through the bottom at a fixed low rate. The amount of coal dust blown out the top was determined and related to coal dustiness. The light-transmittance test apparatus used a helium-neon laser as a light source with a single photo diode as the detector. In this test, coal was dropped down a cylindrical tube and the intensity of the transmitted light was measured at right angles to the tube as the dust settled. The coarse dust index of coal was found to decrease with an increase in moisture content of the coal. The authors also studied other dust suppression agents such as oils. They observed discrepancies between laboratory test and field studies when dust suppressants were used. The authors attributed this disparity to the inefficient application of the dust suppression agents in a field situation.

An indigenous apparatus based on a drop test was used to study factors that affect dust generation (32). A strong dependence of dust generation rate on moisture content was observed for sand and limestone but little change was seen for cement and flour. The dustiness of aluminum oxide and silicon carbide was investigated using the Heubach dustmeter and the MRI dustmeter (26). Moisture content (0.001%–1%) did not affect the size distribution of the dust generated but did decrease the dust generation rate of the powders.

Plinke (33) found that decreasing the moisture content increased dust generation for all materials except dried lactose and glass beads. It was suggested that the decrease could be due to an increase in cohesion values by the formation of solid interparticle bridges when moisture content is reduced in these materials.

Ten powders were used to investigate the role of moisture content on powder dustiness (31). Powders were placed in dessicators that were equilibrated at relative humidities (RH), of 11%, 32%, 53%, 75%, and 98% using saturated salt solutions (34). The powders were tested at 10 L/min. airflow rate and a sample mass of 10 g was used. Two types of powders were encountered with respect to their affinity for moisture, those that absorb greater than 2% moisture and those that adsorb/absorb less than 2% moisture even at very high relative humidities. Avicel PH101, Avicel PH102, croscaramellose sodium, Emcocel 90M, and starch all absorbed greater than 2% moisture and their dustiness indices decreased lin-

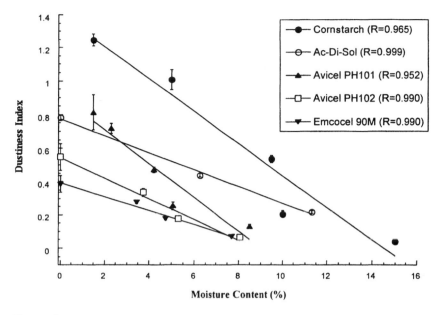

Figure 1 Change in the dustiness index of powders with increase in moisture content. The linear regression coefficient values are: Avicel PH101 (0.95), Avicel PH102 (0.99), Croscaramellose Sodium (0.99), Emcocel 90M (0.99), and Starch (0.97).

early with an increase in moisture content (Figure 1). In Figure 1, the dustiness index of powders is reported only for powders stored in environments up to 75% RH. Powders stored at 98% RH appeared to be in a semisolid state and, therefore, were not tested. Calcium sulfate dihydrate, dicalcium phosphate dihydrate, spray dried lactose, magnesium stearate, and talc adsorbed less than 2% moisture up to 75% RH and the dustiness indices of these powders did not change significantly over the relatively small moisture content changes as compared to their respective dry powder dustiness indices (Figure 2). In general, dustiness indices were highest for powders that were dried at 40° C for 48 hours.

C.   Particle Size and Particle Size Distribution

Particle size governs the generation of airborne particles to a large extent and is the major factor governing the motion of particles in a fluid medium. Therefore, powder particle size and size distribution have the greatest impact on powder dustiness. Generally, an increase in particle size results in a decrease in powder dustiness. Using a fluidized bed device, it was shown that increasing the mean particle size of a powder reduces its dustiness (20).

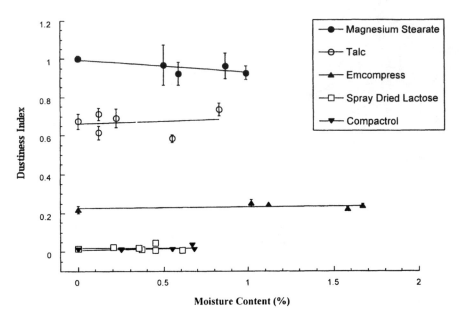

Figure 2   Effect of moisture content on the dustiness index of powders that absorb less than 2% moisture even at high relative humidities up to 70% RH.

With respect to size distribution, the relationship is more complex. Higman (3) observed that the level of dustiness depends upon the proportion of fines present in a bulk material. A large number of samples of coke were specifically blended in various proportions and their dustiness determined in a rolling-drum tester designed by the Warren Spring Laboratory (18). The samples were prepared from three coke powders that had particle size distribution of 250–1000 µm, 50–250 µm, and <50 µm. The maximum dustiness occurred when 15% of material under 50 µm was present.

The dusting potential of alumina was monitored using a modified perra-type dustiness tester (13). The author studied the effects of addition of electro-precipitator dust to alumina powder. The results showed the dustiness index first increases from 0.5 to 2.5 (g dust/kg alumina) with the electroprecipitator dust content up to about 5% and then decreases gradually to 1.0 (g/kg) at 20% fines content. The dustiness behavior of alumina with respect to moisture content was shown to depend upon the thermal treatment of the alumina prior to the test.

Plinke (26) studied the dustiness of aluminum oxide and silicon carbide using the Heubach dustmeter and the MRI dustmeter. An interesting observation by the authors was the phenomenon that dust generated from the "fine material" was not as fine as the parent material. Extracting small particles from a material did not decrease proportionally the amount of dust generated. Materials blended to contain only a portion of small particles were found to be nearly as dusty as materials comprised of entirely small particles. In other words, as the proportion of small particles in a blend diminished, the fraction of these small particles generated as dust increased.

A bench-top apparatus was used to examine factors that affect dust generation (33). This tester employs material impact following a free fall as the mechanism of dust production. The device has a separate measurement section unlike the MRI dustiness tester (15) that consists of only one chamber. The dusty air is pulled at 10 L/min. through an elutriation column that is connected with a five-state impactor for the collection of dust particles according to their aerodynamic size. They hypothesized that the generation rate for particles of a particular size depended upon the fraction of particles of that size and the separation and binding forces of the particles. The binding forces were related to powder-cohesion measurements using a rotational split-level shear tester (35) in which the resisting force of a bulk material was measured against a rotary shear motion. The variable used to describe particle separation forces was impaction, which is the ratio of the impact force divided by material flow. Four different powders were chosen for this study and their dustiness was determined at varying moisture contents and drop heights using the following relationship:

$$G_i = e^{-9.1 \pm 0.96} \times F_i^{-0.2 \pm 0.03} \times \frac{(\text{Impaction})^{1.0 \pm 0.27}}{(\text{Cohesion})^{3.0 \pm 0.11}} \tag{2}$$

where,

G_i = dust generation rate for particles of size "i"

F_i = fraction of particles with size "i" in the test material

$\text{Cohesion} = e^{1.3 \pm 0.56}(M)^{0.2 \pm 0.02}(d_{50})^{-0.2 \pm 0.03}(T_m)^{0.3 \pm 0.08}$

$\text{Impaction} = e^{-1.8 \pm 0.45}(H)^{0.4 \pm 0.1}(M)^{0.1 \pm 0.03}(d_{50})^{0.2 \pm 0.04}(\rho_p)^{0.3 \pm 0.08}$
$(C_w)^{0.2 \pm 0.04}(AR)^{0.3 \pm 0.08}$

where, M is the moisture content of the material, d_{50} is the mass median diameter, T_m is the melting temperature of the material, ρ_p is the particle density, C_w is the width of impact area at its top of the receiving pile, and AR is the angle of repose. The authors concluded if direct measurements of impaction and cohesion can be made, then their method can be used to predict dustiness. However, if equations were to be used to determine impaction and cohesion then the results may be less certain.

Eighteen pharmaceutical powders were studied using the Heubach Dustmeter to determine the relationship between powder dustiness and powder physical characteristics (31). Particle size and size distributions of the powders (test sample) and their dusts were determined by a laser light-scattering device. No size-selective sampling was done because unlike inert powders and minerals, potent pharmaceutical materials may be hazardous when deposited anywhere in the respiratory tract and may cause adverse reactions even by skin contact or ingestion.

Figure 3 is a plot of the dustiness index of dry powders (40° C, 48 hours) versus median particle size of powders that were studied. As seen in the figure, when particle size of the powder is large to begin with, the dustiness of the powder is low. The dustiness increases exponentially for powders in the small particle size ranges. But at lower particle sizes, the dustiness index varies between 0.65 and 1.2 for powders with similar particle sizes. It was concluded that the median particle size of a powder by itself couldn't be used to predict the dustiness of a powder. At lower particle size ranges, other factors like particle shape, density, porosity, and surface area may play a role in the generation of airborne particles from dry bulk powders.

1. Classification of Powders

Pujara (31) compared the particle size distributions of the test sample and the dust sample of powders. The powders fell into one of three categories depending

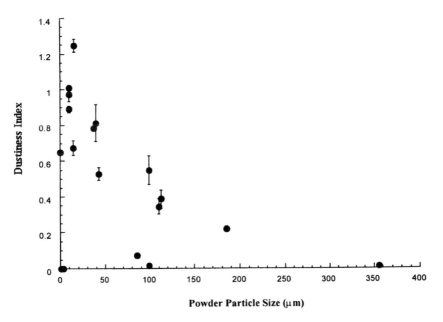

Figure 3 Heubach dustiness index of dry powders as a function of the median particle size of powders determined by a laser light-scattering method.

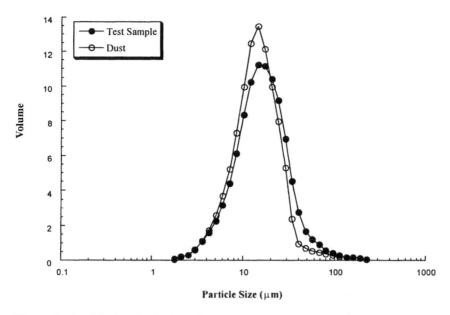

Figure 4 Particle size distribution of magnesium stearate (test sample) and its dust. The test sample is the representative sample of the bulk powder that is placed in the dust generator.

on their dust and test sample size distributions. Figure 4 shows the particle size distributions of magnesium stearate and its dust collected on the filter. The two particle size distributions are very similar and they overlap almost completely. Such powders were classified as Type I powders. Figure 5 shows the particle size distributions of spray-dried lactose and its dust collected in the Heubach Dustmeter. These powders were classified as Type II powders. In this type, the powder-sample size distribution and the dust-sample distribution overlap considerably but the median particle sizes are notably different indicating that only a certain size fraction of the sample selectively becomes airborne and transported to the filter. The median particle size of spray-dried lactose was 100 μm while the median particle size of the dust was 27 μm. Figure 6 shows the particle size distributions of dicalcium phosphate dihydrate and its dust collected using the Heubach Dustmeter. The particle size of the test sample is 185 μm and that of the dust is 8 μm, so the fines from the powder were extracted. These were called Type III powders.

This classification system holds true at an airflow rate of 10 L/min. in the Heubach Dustmeter. At very high flow rates (for example, 100 L/min.), almost all the powders would be Type I since the dust and test-sample size distributions

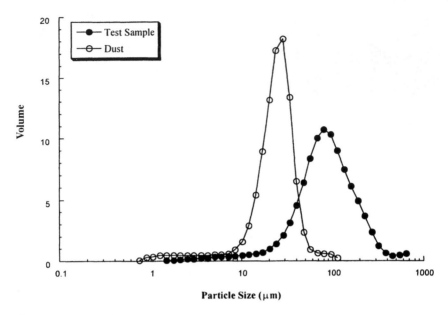

Figure 5   Particle size distribution of spray dried lactose (test sample) and its dust. The test sample is the representative sample of the bulk powder that is placed in the dust generator.

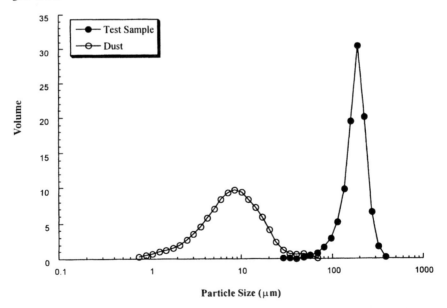

Figure 6   Particle size distribution of dicalcium phosphate dihydrate (test sample) and its dust. The test sample is the representative sample of the bulk powder that is placed in the dust generator.

would be the same for the powders. It was also found that the dustiness index of powders increases with an increase in the percent of particles under 25 μm for Type I, II, and III powders. Thus, the fraction of fines (particle size <25 μm) in a powder is a reasonably good estimate of the dustiness index of powders within a particular class.

2.   Particle Size–Selective Sampling in Dustiness Estimation

A study was conducted to investigate dust generation by free-falling powders (36). The test was basically a drop test done in a vertical chamber with a flow straightener and a high-efficiency particulate air filter. Aluminum oxide powder was dropped onto a flat plate or a beaker of water placed at the bottom of the vertical chamber and the dust-generation process was observed by introducing helium bubbles into the chamber and videotaping the whole process. Particles with an aerodynamic diameter of 20 μm were shown to be present at the top of the chamber even 5 minutes after the powder was dropped. Furthermore, the induced airflow observed with the bubble traces indicated that 40 μm particles can be transported throughout the chamber. The author notes the turbulence and random air motion can be even greater in the industrial environment. Therefore, dustiness tests should not be designed to discriminate against particles of aerodynamic diameter less than 20 to 40 μm.

The particle size–selective sampling criteria adopted by the American Conference of Governmental Industrial Hygienists (ACGIH) for airborne particulate matter, is expressed in three forms (37). The three classifications are inhalable particulate mass, thoracic particulate mass, and respirable particulate mass and are based on particle sizes and their collection efficiencies as discussed in the following text. Inhalable particulate mass consists of those particles captured according to the following collection efficiencies regardless of sampler orientation with respect to air-flow orientation:

$$SI(d) = 50 \% \ (1 + e^{-0.06d}) \quad \text{for } 0 < d \le 100 \ \mu m \tag{3}$$

where, $SI(d)$ is the collection efficiencies for particles with aerodynamic diameter d in μm.

For thoracic particulate mass:

$$ST(d) = SI(d)[1 - F(x)] \tag{4}$$

where, $x = \dfrac{\ln(d/\Gamma)}{\ln(\Sigma)}$

$\Gamma = 11.64$ μm where gamma is chosen such that for a particle aerodynamic diameter of 10 μm, 50% of the ambient particles are included in the thoracic fraction of the inhalable particulate mass.

Sigma = 1.5 where sigma is a curve fitting parameter that is related to the steepness of the curve (thoracic particulate mass versus particle aerodynamic diameter).

$F(x)$ = the cumulative probability function of a standardized normal variable, x. Finally, for respirable particulate mass:

$$SR(d) = SI(d)[1 - F(x)] \qquad (5)$$

where, $F(x)$ has the same meaning as previous with:

$\Gamma = 4.25 \ \mu m$, gamma chosen such that for a particle aerodynamic diameter of 4 μm, 50% of the ambient particles are included in the respirable fraction of the inhalable particulate mass.

$S = 1.5$, sigma being a curve-fitting parameter related to the steepness of the curve (respirable particulate mass versus particle aerodynamic diameter).

For a detailed description and history of gamma and sigma, please refer to S.C. Soderholm (1989).

Potent pharmaceutical materials may be hazardous when deposited anywhere in the respiratory tract defined previously by the inhalable particulate mass category. Therefore, dustiness studies of potent pharmaceuticals should not be discriminated against particles of aerodynamic diameter less than 10 to 100 μm.

D. True Density

By itself the true density of powders cannot affect powder dustiness because particle size is the most important parameter governing the motion of aerosol particles. But, along with particle size, density affects the aerodynamics of the particle since particle density determines the mass of a particle.

Alumina dustiness was studied using a Perra pulvimeter (11) and also a fluidization unit to compare the two methods (21). Both methods proved to be useful in estimating alumina dustiness qualitatively. Calciner discharge (electrostatic precipitator dust) was added to alumina samples in various proportions (2%–12%). Dustiness was found to increase with increasing calciner dust, which is different from the results obtained by Hsieh (13). No clear relationship was observed between density and dustiness for this particular powder.

E. Flowability/Powder Cohesion

The angle-of-repose technique has been used to determine powder flowability and cohesion. This technique has some disadvantages that make it subjective. Determining the "right" angle is difficult because the height of the cone is not stable for many cohesive materials. Also, sieving a material onto a plate may tend to classify the material by size so that the cone material does not have the same size distribution as the original material. A few studies have been done to correlate the angle of repose with powder dustiness.

The angle of repose correlated poorly with the dustiness index when a modified Perra pulvimeter (13) was used to study alumina dustiness (30).

Cowherd (15) investigated the use of the angle of repose to predict dustiness in an industrial setting. They speculated that the steeper the angle of repose, the lower the dust generation. Angle of repose was measured for 14 materials each of which had a different moisture content and particle size distribution. However, they found little correlation between dust generation rate and angle of repose.

The effect of powder cohesion on dust generation was also evaluated using a bench-top apparatus (10). The cohesiveness of a powder was determined using a shear tester and was related to the powder moisture content, melting temperature, and mass median size of the particles. Powder cohesion was described by the following equation,

$$\text{Cohesion} = e^{1.3 \pm 0.56}(M)^{0.2 \pm 0.02}(d_{50})^{-0.2 \pm 0.03}(T_m)^{0.3 \pm 0.08} \tag{6}$$

where, M is the moisture content of the material, d_{50} is the mass median diameter, and T_m is the melting temperature of the material. It was found that dustiness decreased with an increase in cohesiveness of the powder. Shear testers have been used to determine powder cohesion when related to compaction properties of powders. Using a shear tester to determine cohesion in powders for aerosolization behavior of powders is not appropriate because in a shear tester, the shear stress of powders is measured at different normal stresses that are not encountered in the air streamlines. Although shearing action is required to aerosolize powders, the forces are much smaller and the powder is not deformed in any manner unlike the shear tester due to the load placed on the powder bed. Questions about the reproducibility of cohesion measurements obtained using shear testers sill exist. Also, shear testers have the disadvantage of being subjective tests when used to measure powder cohesion. Powder cohesion is a property that is dependent upon many physical characteristics of powders and attempts have been made to measure cohesion using other techniques (38–40) but to date no single instrument has proved to be satisfactory in determining powder cohesion as it relates to powder aerosolization.

F. Effect of Particle Shape on Powder Dustiness

The variation between diameters obtained by sizing instruments increases as the particles diverge from a spherical shape. Hence shape is an important factor in the correlation of sizing analyses made by various procedures. Many powder properties depend on particle shape (41–45). Consequently, it needs to be addressed in systems where shape is suspected to play a role, which may be the case with powder dustiness.

Authier-Martin (30) studied the effects of alumina bulk properties on its dustiness. No correlation was found between bulk properties of alumina and its dustiness. A fair correlation was found between the median size of the dusts and the dustiness index of the bulk material. That is, the dustiness index decreased with increase in median size. Scanning electron micrographs showed a difference between the dusty and nondusty alumina samples. This led the author to speculate on particle shape as being a critical parameter in dustiness of alumina. No quantitative studies on particle shape were performed and, overall, no analysis was done on the effect of particle shape on dustiness of alumina.

Particle shape of eighteen powders were determined using an image analysis system (31). These shape factors (aspect ratio, circularity, elongation ratio, roundness, and sphericity) by themselves did not correlate well with the dustiness indices of powders. Also, these shape factors when multiplied by the median particle size, did not correlate well with the dustiness indices of powders.

The surface-volume shape coefficient ($\alpha_{sv,a}$) has been reported in the literature to describe several powder properties and effects due to shape (46–49). For example, a form of $\alpha_{sv,a}$ has been used to describe the inhalation properties of two disodium cromoglycate powders (50).

$$\alpha_{sv,a} = S_w \times \rho_p \times d_{sv} \qquad (7)$$

where, S_w is the specific surface area (m^2/g), ρ_p is the true density (g/cm^3) of the powder and d_{sv} is the surface-volume mean diameter. d_{sv} is an average size based on the specific surface-per-unit volume.

$$d_{sv} = \frac{\Sigma d_a^3}{\Sigma d_a^2}$$

where d_a is the area projected diameter (μm).

Wong and Pilpel (51) described the effect of particle shape on the mechanical properties of powders by modifying the above equation ($\alpha_{sv,a}$) in the following manner:

$$\text{Shape Coefficient} = (\text{Heywood Equivalent Diameter} \times S_w \times \rho_p) \qquad (8)$$
$$+ \text{Elongation Ratio}$$

where

$$\text{Heywood Equivalent Diameter} = \sqrt{\text{Maximum Diameter} \times \text{Minimum Diameter}}$$

$$\text{Elongation Ratio} = \frac{\text{Maximum Diameter}}{\text{Minimum Diameter}}$$

Since size plays a major role in powder aerosolization behavior, this shape coefficient was studied along with the median particle size (shape coefficient $\times$ median particle size) to determine its relationship with powder dustiness. When the dustiness index was plotted against the previously mentioned parameter, powders with elongation ratios greater than 2.0 such as acetaminophen, deviated from the linear regression line resulting in a poor correlation (31). A factor overlooked in the previous equation was the use of elongation ratio by Wong and Pilpel (51). The powders studied by Wong and Pilpel (51), did not have a large elongation ratio; in fact, the maximum elongation ratio observed was 1.31. Pujara (31) studied powders such as acetaminophen (Elongation Ratio = 4) and other powders that have elongation ratios greater than 2. Therefore, a simple alteration of Equations 7 and 8 resulted in the following shape coefficient (K_p):

$$K_p = d_{sv} \times S_w \times \rho_p \times \text{Elongation Ratio} \tag{9}$$

The surface-volume diameter (d_{sv}) was used instead of the mean Heywood equivalent diameter because the mean Heywood equilvalent diameter can be used only if the size range is narrow and the distribution is normal. These conditions are rarely found in pharmaceutical powders (52, 53).

The empirical shape coefficient, K_p, when multiplied with median particle size of powders, correlated linearly with the powder dustiness index (31). (See Figure 7). Sodium starch glycolate however (dustiness index = 0.53) deviates substantially from the regression line. The reason for this is not clear.

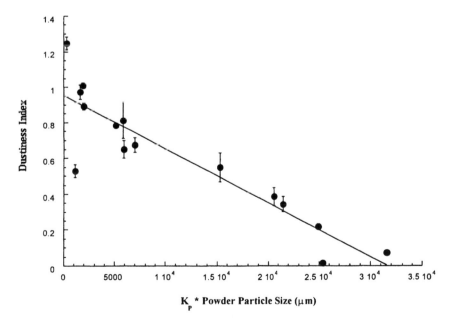

Figure 7 Dependence of dry powder dustiness on particle size and shape where K_P is a powder shape coefficient ($r = 0.91$).

G. Summary

Dustiness may be related to more than one physical parameter of a powder. Among all the powder parameters discussed it this section, particle size has the greatest effect on powder dustiness. When powders have similar particles sizes, then particle size distributions, shape, surface area, density and cohesion become important with respect to dustiness. The need to identify the critical parameters of powders that contribute to powder dustiness still exists and is of increasing importance due to the emergence of expensive and extremely potent drugs in the pharmaceutical industry.

V. DUST IN THE PHARMACEUTICAL INDUSTRY

Dust of any kind in a pharmaceutical facility can be a serious problem. Cross-contamination of products can occur in the industry if dust levels are not monitored and maintained to a minimum level. For example, cross-contamination of a product with penicillin can cause deleterious effects to a person sensitive to penicillin.

Product loss due to dust emissions can result in significant increases in cost of manufacturing of expensive drugs. Gold et al. (5) determined dustiness of eleven powders commonly used in the pharmaceutical industry. The results were used to classify powders into three categories according to their dust counts. No further analysis of the data was done.

Workers in the pharmaceutical industry may also be exposed to compounds designed to produce extreme pharmacological effects. Workers suffered with airborne contact urticaria due to sodium benzoate in a pharmaceutical manufacturing plant (54). Immunologic reactions and modification of normal intestinal bacterial flora have been reported in penicillin factory workers (55, 56). Workers in a factory manufacturing cimetidine tablets developed respiratory symptoms related to periods at work (57). There are many more examples where workers have been exposed to potent dust in the pharmaceutical industry (58–61). Sargent and Kirk (62) noted that many pharmaceutical compounds that have been manufactured for several decades, have no guidelines for safe exposure levels. But significant measures have been taken by the pharmaceutical industry to control the exposure of production workers to pharmacologically active health products.

The first step for a proper health risk assessment consists of establishing adequate exposure limits for each drug of interest. To accomplish this task, knowledge about the dustiness of a powder during a handling should be known. When possible, this would also help the production plant to control dustiness by using powders with inherently low dustiness. Increasingly, product development scientists have the added responsibility of preparing products that generate minimal dust. Unfortunately, not much information is available on the selection of low dust-yielding ingredients. Dustiness studies on pharmaceutical powders are very sparse in the literature. Therefore, studies to determine dustiness of pharmaceutical powders need to be conducted and methods to facilitate routine evaluations of powders must be discovered.

APPENDIX: HEUBACH DUSTMETER

The Heubach Dustmeter (Heucotech Ltd., Fairless Hills, PA) is one of the most commonly used instruments to determine powder dustiness. The instrument consists of a horizontal rotating drum with internal baffles that produces a repeated powder fall through a regulated airstream (Figure 8). A weighted amount of powder is placed in the dust-generating drum that rotates at 30 rpm. The dust generator has three baffles at an angle of 45° to the case wall that repeatedly lift the powder and the powder free falls through about 6 cm to the bottom of the generator. The dust generator has a diameter of 14 cm so the powder does not fall into the core of the airstream. Therefore, the dust is generated from the turbulence created by the falling powder and by the powder that impacts the bottom of the drum. A vacuum pump is used to draw air at a fixed flow rate through the shaft of the rotating drum containing the powder. The air transports the dust generated in the drum through the settling chamber. The airborne particles can be subsequenty evaluated by a variety of methods. Usually, a 30-second or 1-minute time period is sufficient for collection of the dust. The settling chamber, also known as the coarse-particle separator is used to separate out large particles that may bounce out of the rotating drum and particles that are not truly entrained in the air. The weight of the dust may be determined by collecting it onto a preweighted filter. The Heubach dustiness index can then be calculated using the following formula:

$$\text{Dustiness Index (g/g min)} = \frac{\text{Dust collected on the filter (g)}}{\text{Mass of material tested (g)} \times \text{Test Period}} \times 100$$

(10)

where, the Test Period is the length of time that the dust is collected.

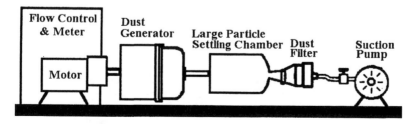

Figure 8   The Heubach Dustmeter (Heucotech Ltd., Fairless Hills, PA).

REFERENCES

1. L Svarovsky. Powder Testing Guide: Methods of measuring the physical properties of bulk powders. New York: Elsevier Science Inc., 1987, pp 122.
2. C Schofield. Dust generation and control in materials handling. Bulk Solids Handl. 1(3):419–427, 1981.
3. RW Higman. Dustiness Testing: A useful tool. In: HD Goodfellow, ed. Ventilation '85. New York: Elsevier Science, Inc., 1986, pp 693–702.
4. HD Goodfellow, JW Smith. Dustiness Testing: A new approach for dust control. In: JH Vincent, ed. Ventilation '88. Elsevier Science, Inc., 1989, pp 175–182.
5. G Gold, RN Duvall, BT Palermo, RL Hurtle. Dustiness of pharmaceutical formulations I: Instrumentation. J. Pharm. Sci. 62(9):1530–1533, 1973.
6. ASTM D547. Standard method of test for index of dustiness of coal and coke. Annual Book of ASTM Standards. Part 26: 239–242, 1975.
7. RJ Mikula, IS Parsons. Coal dustiness: characterization and control. Coal Preparation, 9:199–212, 1991.
8. AB Wells, DJ Alexander. A method for estimating the dust yield of powders. Powder Tech. 19:271–277, 1978.
9. SL Sutter, JW Johnston, J Mishima. Investigation of accident-generated aerosols: releases from free fall spills. Am. Ind. Hyg. Assoc. J. 43(7):540–543, 1982.
10. B Cawley and D Leith. Bench-top apparatus to examine factors that affect dust generation. Appl. Occup. Environ. Hyg. 8(7):624–631, 1993.
11. S Perra. Measurement of sandy alumina dustiness. Light Metals 1984 pp 269–286, 1984.
12. DA Lundgren. A measurement technique to quantitate fugitive dust emission from handling of granular products. J. Aerosol Sci. 17(3):632–634, 1986.
13. HP Hseih. Measurement of flowability and dustiness of alumina. Light Metals 1987. Proceedings of American Institute of Mining, Metallurgical and Petroleum Engineers (AIME) Annual Meeting, Metallurgical Soc. of AIME, Warrendale, PA, 1987, pp 139–149.
14. KH Carlson, DR Herman, TF Markev, RK Wolff, MA Dorato. A comparison of two dustiness evaluation methods. Am. Ind. Hyg. Assoc. J. 53(7):448–454, 1992.
15. C Cowherd Jr., MA Grelinger, PJ Englehart, RF Kent KF Wong. An apparatus and methodology for predicting the dustiness of materials. Am. Ind. Hyg. Assoc. J 50(3): 123–130, 1989.
16. D Stauber, R Beutel. Determination and control of the dusting potential of feed premixes. Fresenius Z. Anal. Chem. 318:522–524, 1984.
17. TR Farrugia, N Ahmed, GJ Jameson. A new technique for measuring the dustiness of coal. J. Coal Quality. 8(2):51–55, 1989.
18. British Occupational Hygiene Society. Dustiness estimation methods for dry materials: Part 1, Their uses and standardization and Part 2, Towards a standard method (Technical Guide No. 4). Norwood, Middlesex, England, Sciences Reviews Ltd., 1985.
19. W Cator, A Gray. Evaluating the dustiness of powders. Powder Handl. Process. 2(2): 145–148, 1990.
20. C Schofield. The generation of dust by materials handling operations. J. Powder & Bulk Solids Tech. 3(1):40–44, 1979.

21. AJ Chambers. Assessment of alumina dustiness. Powder Handl. Process. 4(1):47–52, 1992.
22. KYK Chung, GJ Burdett. Dustiness testing and moving towards a biologically relevant dustiness index. Ann. Occup. Hyg. 38(6):945–949, 1994.
23. WA Heitbrink, WF Todd, TJ Fischbach. Correlation of tests for material dustiness with worker exposure from the bagging of powders. Appl. Ind. Hyg. 4(1):12–16, 1989.
24. WA Heitbrink. Factors affecting the Heubach and MRI dustiness tests. Am. Ind. Hyg. Assoc. J. 51(4):210–216, 1990.
25. WA Heitbrink, WF Todd, TC Cooper, DM O'Brien. The application of dustiness tests to the prediction of worker dust exposure. Am. Ind. Hyg. Assoc. J. 51(4):217–223, 1990.
26. MAE Plinke, R Maus, D Leith. Experimental examination of factors that affect dust generation by using Heubach and MRI testers. Am. Ind. Hyg. Assoc. J. 53(5):325–330, 1992.
27. P De Keersmaeker. Council Directive, The Official Journal of the European Communities. OJ No L 160, 20. 6:32–35, 1987.
28. F Andriessen. Commission Directive, The Official Journal of the European Communities. OJ No L. 11, 14. 1:34–35, 1989.
29. KM Davies, CM Hammond, RW Higman, and AB Wells. Progress in dustiness estimation. Ann. Occup. Hyg. 32(4):535–544, 1988.
30. MM Authier. Alumina Handling Dustiness. Light Metals 1989. Proceedings of sessions, American Institute of Mining, Metallurgical and Petroleum Engineers (AIME) Annual Meeting, Metallurgical Soc. of AIME, Warrendale, PA 1989, pp 103–111.
31. C Pujara. Determination of factors that affect the generation of airborne particles from bulk pharmaceutical powders. Ph.D. Dissertation, Purdue University, West Lafayette, IN, 1997.
32. MAE Plinke, D Leith, DB Holstein, MG Boundy. Experimental examination of factors that affect dust generation. Am. Ind. Hyg. Assoc. J. 52(12):521–528, 1991.
33. MAE Plinke, D Leith, MG Boundy, F Loffler. Dust generation from handling powders in industry. Am. Ind. Hyg. Assoc. J. 56(3):251–257, 1995.
34. H Nyqvist. Saturated salt solutions for maintaining specified relative humidities. Int. J. Pharm. & Prod. Mfr. 4(2):47–48, 1983.
35. IASZ Peschl. Equipment for the measurement of mechanical properties of bulk materials. Powder Handl. Process. 1(1):73–81, 1989.
36. WA Heitbrink, PA Baron, K Willeke. An investigation of dust generation by free falling powders. Am. Ind. Hyg. Assoc. J. 53(10)617–624, 1992.
37. American Conference of Governmental Industrial Hygienists, Threshold limit values for chemical substances and physical agents. Technical Information Service, Cincinnati, OH, 1992, pp 41–43.
38. RL Carr. Evaluating flow properties of solids. Chem. Engng. 163–168, 1965.
39. HH Huasner. Friction conditions in a mass of metal powder. Int. J. Powder Metallurgy. 3(4):7–13, 1967.
40. PL Bransby. Current work in materials handling at Warren Spring Laboratory. The Chemical Engineer 3:161–164, 1977.

41. W Stober. Dynamic shape factors of nonspherical aerosol particles. In: T Mercer, ed. Assessment of Airborne Particles: Fundamentals, Applications and Implications of Inhalation Toxicity. Proceedings, Rochester International Conference on Environmental Toxicity (3rd: 1970), T. Mercer, P. Morrow, W. Stobev, Eds., Thomas, Springfield, IL (1972), pp 249–289.

42. M Bergeron, P Laurin, R Tawashi. Effects of particle morphology in selecting pharmaceutical excipients. Drug Dev. Ind. Pharm. 12:915–926, 1986.

43. JK Beddow, AF Vitter, K Sisson. Powder Metallurgy Review 9, Part I: Particle shape analysis. Powder Metallurgy International. 8(2):69, 70, 75, 76, 1976.

44. I Gonda, AF Khalik. On the calculation of aerodynamic diameters of fibers. Aerosol Sci Tech 4:233–238, 1985.

45. L Hellen, J Yliruusi. Process variables of instant granulator and spheroniser: III. Shape and shape distribution of pellets. Int. J. Pharm. 96:217–223, 1993.

46. LW Wong, N Pilpel. The effect of particle shape on the mechanical properties of powders. Int. J. Pharm. 59:145–154, 1990.

47. CA Walton, N Pilpel. The effects of particle size, shape and moisture content on the tensile properties of procaine penicillin powders. J Pharm Pharmacol, 24:10P–16P, 1972.

48. I Nikolakakis, N Pilpel. Effects of particle shape and size on the tensile strengths of powders. Powder Tech. 56:95–103, 1988.

49. T Pesonen, P Paronen. Evaluation of a new cellulose material as binding agent for direct compression of tablets. Drug. Dev. Ind. Pharm. 12(11–13):2091–2111, 1986.

50. MT Vidgren, PA Vidgren, TP Paronen. Comparison of physical and inhalation properties of spray dried and mechanically micronized disodium cromoglycate. Int. J. Pharm. 35:139–144, 1987.

51. LW Wong, N Pilpel. Effect of particle shape on the mixing of powders. J. Pharm. Pharmacol. 42:1–6, 1990.

52. IC Edmundson. Advances in Pharmaceutical Sciences, Vol. 2. HS Bean, JE Carless, AH Beckett, Eds. London: Academic Press, 1967, p 95.

53. A Martin, J Swarbrick, A Cammarata. Physical Pharmacy: Physical Chemical Principles in the Pharmaceutical Sciences. Philadelphia: Lea & Febiger Co., 1983, pp 506–507.

54. JR Nethercott, MJ Lawrence, A Roy, BL Gibson. Airborne contact urticaria due to sodium benzoate in a pharmaceutical manufacturing plant. J. Occup. Med. 26(10): 734–736, 1984.

55. E Shmunes, JS Taylor, LD Petz, G Garatty, HH Fudenberg. Immunologic reactions in penicillin factory workers. Ann Allergy. 36:313–323, 1976.

56. FL Vil'Sanskaja, GB Steinberg. Modification of the bacteria of the intestine and other organs following occupational exposure to antibiotics (Streptomycin, Tetracycline, and Penicillin). Gig. Tr. Prof. Zabol. 14:25–31, 1970.

57. II Coutts, S Lozewicz, MB Dally, AJ Newman-Taylor, P Sherwood Burge, AC Flind, DJH Rogers. Respiratory symptoms related to work in factory manufacturing cimetidine tablets. British. Med. J. 288:1418, 1984.

58. RW Newton, MCK Browning, J Iqbal, N Piercy, DG Adamson. Adenocortical suppression in workers manufacturing synthetic glucocorticoids. Br. Med. J. 1:73–74, 1978.

59. VDM Stejskal, R Olin, M Forsbeck. In: AW Hayes, RC Schnell, TS Miya, eds. Diagnosis of drug induced occupational allergy by lymphocyte transformation test. Developments in the Science and Practice of Toxicology. New York: Elsevier Science, Inc., 1983, pp 559–562.
60. L Ekenvall, M Forsbeck. Contact eczema produced by a β-adrenergic blocking agent (Alprenolol). Contact Dermatitis 4:190–194, 1978.
61. RA hardie, JA Savin, DA White, S Pumford. Quinine dermatitis: Investigation of a factory outbreak. Contact Dermatitis 4:121–124, 1978.
62. EV Sargent, DG Kirk. Estblishing airborne exposure control limits in the pharmaceutical industry. Am. Ind. Hyg. Assoc. J. 49(6):309–313, 1988.

5

Particle Monitoring: Old and New

Brian G. Ward
Eli Lilly and Company, Indianapolis, Indiana

I. INTRODUCTION

At this time, it is very costly and time consuming to perform the level of conventional sampling and laboratory analysis needed for compounds having exposure guidelines in the ng/m^3 concentrations. The pharmaceutical industry is at frequently beyond the limiting edge of current analytical chemical specia tion techniques. Air sampling times can become extremely long at these levels, and incapable of revealing transient particulate releases in the workplace. For a manufacturing facility to control its destiny, the sampling and data analysis (and its frequency) and sampling strategy have to be enhanced to provide regular feedback on facility performance. Ideally, feedback should be almost instantaneous to afford the maximum advantage to operations.

Optical particle-counting instruments hold great promise for use in identifying and measuring releases of high-potency compounds. Optical particle counting based on discrete particle counts lends itself to observing the ambient versus incident experiences at very low levels of fugitive airborne releases. These instruments are conventionally used as total particle counters to measure compliance with International Standards Organization (ISO) standard 14644 (having evolved from ISO 9000 and 9002) for clean rooms. By contrast, conventional industrial hygiene optical particle counters (OPCs) are designed for use in a relatively contaminated background where typical exposure levels of the $\mu g/m^3$ concentrations and upward are encountered.

II. BACKGROUND

In the continuing development of potent drugs with associated exposure guidelines in the ng/m^3 range, the industry is gravitating toward a work environment
in which the process is contained, providing minimal fugitive releases. Traditionally, manufacturing relied on the use of personal protective equipment (PPE),
providing an environment where fugitive releases are expected to occur. There
are strong financial drivers to keep the product in place—including product cost,
downtime for cleanup, and quality, environmental, and occupational health considerations. As a result of this initiative of contained processing, new containment
concepts have been developed to mitigate fugitive releases into the processing
suites. This has placed extreme pressure on the ability of the analytical laboratory
to provide finite measurement of releases of active compounds. Ambient background contaminant levels in the workplace progressively have been cleaned up
such that optical particle counters, such as the CLiMET, Grimm, Hitachi, MET,
and related instruments are still within their operating range without overloading
the counter. Use of clean room instruments is not feasible in a heavily contaminated work area since it is easy to saturate their sensing capacity. Measurably
high ambient backgrounds in the relative particle-size range of these instruments
do not render them useless as long as background counts are stable, may be
tracked, or are at least predictable. The devices are best used to measure transient
change excursions and to associate the change with a particular operational procedure or incident. The availability of real-time data enables the operating department to remediate fugitive releases through both engineering and procedural
changes and to observe the consequence, all within a short period of time—often
minutes.

It is important to not overlook the additional benefit of the operators becoming comfortable with the data and its meaning, which requires that feedback is
needed whenever an improvement or degradation of hardware, software, or operational procedures has occurred. In many applications, significant operator buy-
in to the concept of contained processing can be achieved through experiences
gained with operator collaboration and real-time data feedback from the sampling
practitioner. This operator advantage cannot be overstated. The best engineering
is only as good as the operating procedures accompanying it and the diligence
of the operator involved.

The issue of particle size has also become of great significance since bio-
availability pathways are important in relation to exposure guidelines for active
compounds. Respiratory deposition profiles (i.e., tracheal, bronchial, and alveolar-
deep lung) lead to enhanced need for size-specific data. This leads to ever-greater
demands for analytical or sensing systems since this very selectivity leads to
decreased particle masses collected.

The issue of the aerodynamic behavior of particles ("aerodynamic particle size") and the optical equivalent particle size is a very real one in this environment, as aerodynamic sampling becomes less capable of delivering data.

III. CONCEPTS AND DEFINITIONS

To fully understand particle-counting methods, it is important to have a clear understanding of key concepts, including:

Different ways of measuring particle size
Nist* standard procedures for calibrating instruments
Typical clean room concepts that may be applied in the pharmaceutical industry
Appropriate placebo-identification factors such as dustiness
Analytical techniques used in particle counting

A. Particle Size

Table 1 describes the two traditional ways particle size is measured: optically and aerodynamically.

B. Calibration of Air-Sampling Devices

This section provides detailed information of NIST standards for calibrating various air-sampling devices. It also points out where no such standards exist and how, as a reference, one pharmaceutical company has performed calibration. The active calibration of a single sampling inlet should not in any way affect the sampling pump and attached system. The calibration device should introduce no pressure drop, but in the nonideal world the least pressure differential is probably introduced by cautious use of a soap-film flowmeter. Ways to circumvent the problems described here include calibrating the soap-film flowmeters with a secondary standard rotameter using stable airflow or using a limiting orifice with appropriate critical pressure drop across the orifice in the flow range of normal calibrations. The ideal situation for calibration is to use a displacement spirometer with continual balancing to ensure that the chamber is at atmospheric pressure as the spirometer volume is displaced. This will ensure a full traceable NIST

---

* National Institute for Science and Technology (formerly the National Bureau of Standards).

Table 1 Particle Size Measurement Options

Particle size type	Calibration standard	Assumptions	Data output
Optical	Standard polystyrene spheres (NIST traceable)	• Isokinetic entrance velocity • Spherical particle equivalency • Comparable optical reflectance • Known particle density (or factor)	• Optical particle counts by size range (sometimes) • Mass equivalent concentration (actual or computer macro manipulation) • Data manipulatable for background correction
	Arizona road dust (NIST traceable)	• Assumes for calculation that each sample is Optically equivalent to the particle-size distribution profile of arizona road dust	• A single mass/m^3 value
Aerodynamic	Flow rate prediction of particle-size cut point	• Isokinetic entrance velocity • True aerodynamic differentiation through inertial impaction • Density dependency • Shape equivalency • 100% size-segregation efficacy (actually 50%)	• Mass concentration by particle size range • Chemical assay concentrations by particle-size range are feasible

calibration. However, it is typically not feasible in the workplace due to the sheer bulk and skill required in using such a system.

1. Soap-Film Flowmeters

Commercial soap-film flowmeters or a laboratory burette are not NIST calibrated standards. NIST traceability implicitly requires that the calibrating device be calibrated in the same manner in which it is used in practice. There are no existing NIST procedures for calibration of these devices. All claims for NIST traceability are based on calibration by the volume displacement or equivalent method and not true end-use conditions. NIST has declined to consider a calibration certification for soap-film flowmeters due to many factors, including soap-film concentration density and viscosity; temperature corrections for both the liquid film and the glass chamber; and consistency of generating the bubble. Specific calibration and performance ratings should be requested from the manufacturer. Figure 1 shows the calibration profile of a typical soap-film flowmeter. The singular advantage of the soap-film flowmeter is the absence of significant pressure drop (<0.1 in. water gage) during use, minimizing the effect on the pump's delivery capacity.

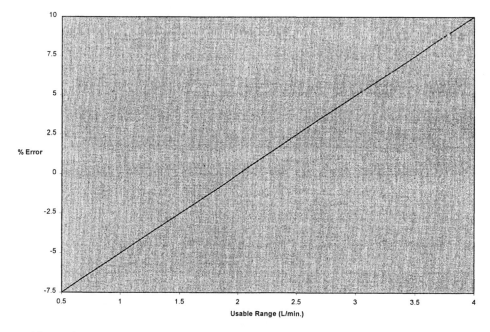

Figure 1   Typical soap-film flowmeter calibration profile.

2. Rotameters

Single-ball rotameters are secondary standard NIST calibratable flow-measuring devices. Use of two-ball rotameters is not subject to the same NIST calibration standard. Rotameters are calibrated relying on the fact that the pressure at the entrance to the rotameter is at barometric condition and that the downstream pressure is also traceable. Double ball rotameters do not provide the means for measuring the downstream pressure for the first ball and simultaneously the upstream pressure for the second ball. Consequently, standard correction factors do not apply to two-ball rotameters. An addiitonal problem using a rotameter is that the mass and flow restrictions due to the ball floating inside the tapered tube of the rotameter can directly affect the volume of airflow flowing through the pump due to mechanical factors, e.g., flow-adjusting pumps. It becomes obvious that the use of flowmeters with installed needle valves causes ever-greater pressure-drop issues.

3. Floating Metal Bob Flowmeters

Similar in concept to the soap-film flowmeter, floating metal bob flowmeters differ in use because they create a pressure drop that will in turn affect the flow rate of the pump during calibration. The problem of differential pressure changes affecting the sampling pump is further exaggerated using the multiple sampler heads provided by some vendors. These heads function in a restrictive orifice mode rather than a limiting orifice mode. As a result, any change of pressure at a single inlet point will affect the proportional flow through all of the other sampling devices and orifices attached to the head. This occurs whenever a flowmeter is attached to any one of the heads thereby introducing a pressure drop.

4. Sampling Pumps

Typical battery-operated sampling pumps may display an average flow rate of, for example, 2 L/min. The maximum and minimum range in which this pump actually functions can be as wide as 0.5 L/min. to 4 L/min. with cyclical fluctuations due to the duty cycle of the nonreturn valves that control the pump's action as illustrated in Figure 2. The actual flow rate through the sampling device can be of lesser dynamic cycling range purely due to the differential pressure introduced to the sampling train by the sampling medium and associated devices. The cost for reduced flow fluctuations is reduced run time due to battery drainage. Recent practices (e.g., the asbestos abatement industry) have introduced flexible Luer fittings for attaching the sampling cassette to the sample hose. These in turn are compressible and, especially with repeated use, will deform to create orifice-like openings in the sampling manifold (described in the Specialized Monitoring Techniques section) that further reduce the capacity of the pump but improve

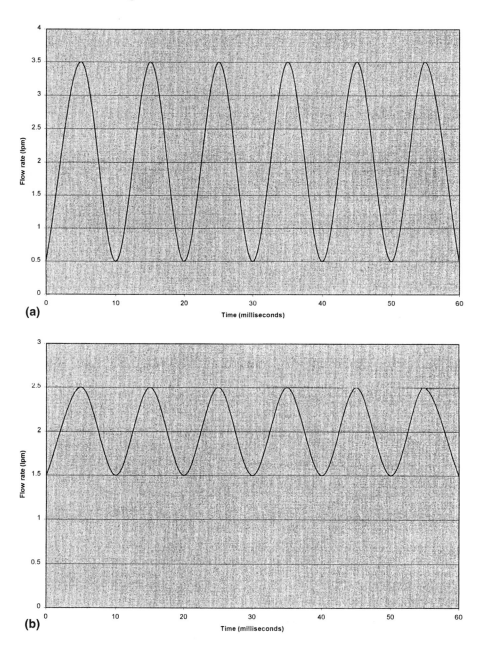

Figure 2 Cyclical fluctuations in battery-operated pump at average flow rate (e.g., 2 L/min.): (a) with no pressure drop at the inlet; (b) with low pressure drop at the inlet.

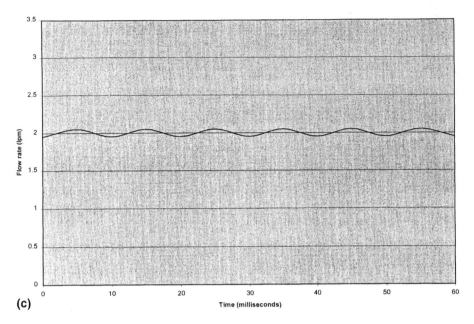

(c)

Figure 2   Continued. (c) With pressure drop ~ halfway into the pump range.

the duty cycle. As they become more flexible with use, they become too flexible and create too small an orifice for the pump to function. Sample hose, sample hose connectors, sampling devices, and sampling media must be considered seriously as part of the overall calibration system. The use of robust sampling pumps with proven self-adjusting flow characteristics can minimize the effects of such devices, but have to be frequently checked to ensure that the pump system parts are fully functional.

Whenever possible, introduce a sampling manifold, for example at fixed sampling locations, to deliver an adequate air-sampling flow to a limiting orifice. In other words, one that is working at sonic velocity, which by the laws of physics implicitly defines the maximum stable flow rate through the orifice. This provides total flow control and flow reproducibility when used in conjunction with a vacuum pump having a negative pressure gauge or indicator, and in the case of rotary vane pumps, a pressure relief valve to prevent pump starvation. It is the responsibility of the practitioner to ensure that the sampling medium is not of sufficient microporous behavior as to overwhelm the sonic performance of the limiting flow orifice. Overwhelming the limiting orifice by actions such as creating a high-pressure drop across the sampling head will result in inadequate performance of the sample train due to heavy vacuum load and inability to provide sufficient negative pressure and flow. Examples of such sampling media are Polytetrafluoroethylene (PTFE) microporous membranes that will result in a limiting orifice capable of acting only as a restrictive orifice. In other words, subsonic velocity through the orifice has less direct control whenever adjacent sampling heads show change in flow (just as a multiple sampling head).

5. Battery-Operated High-Flow Pumps

Battery-operated high-flow pumps are notorious in their inability to maintain constant flow even over short time duration, due to flow following an exponential pressure decay. In addition, they have limited ability to perform under negative pressure (e.g., minus 12 in. H_2O gage). Current calibration procedures can readily be observed to affect the actual flow through the pump and sampling device, resulting in imposed upstream pressure drops and artificial calibration data. The pressure drop between sampling device and flowmeter should be extremely low (<0.2 in. H_2O gage) in these calibration scenarios. Most high-flow pumps use a built-in flowmeter with needle valve in-line for flow measurement. The flowmeter is not at barometric pressure and can require serious flowrate correction in use.

C. Clean Room Air-Monitoring Criteria

By definition, a clean room must have a dust free environment. Typical dust monitoring is performed by counting particles down to either 0.5 μm diameter or 0.3 μm diameter in size, depending on the standard, to show the performance of the entrance air High Efficiency Particulate Air filter (HEPA) filtration efficacy and the gowning, robing, and cleanup procedures. As a result, many of the commercially available particle counters have been programmed to provide data output based on these standards (e.g., ISO 14644 criteria). The rationale for this is that only the small particles have the potential for staying airborne for significant lengths of time. The larger particles, being rather like rocks, have a rapid settling velocity with limited horizontal transport across distances. In these environments, isokinetic sampling (i.e., sample intake velocity matches the ambient airflow velocity) is used, but frequently does not detect large counts of the larger particles. Monitoring devics in a clean room are used effectively to measure a "go/no go" situation. The utility of this instrument, however, is not limited to clean room monitoring.

D. Placebo Selection

It is all too frequent that the cost of the raw materials to be handled in the workplace is high and the toxicity leads to extreme care in handling. As a result, it is necessary to perform equipment and procedural testing using a placebo. The placebo selected must be a material having known behavior characteristics and a sensitive assay, but of low toxicity. After extensive studies, a limited number of materials have been shown to be effective surrogates for performance testing based on particle size, density, static charge, and feasible chemical analysis in the range of high-potency materials. For example, lactose assays have been developed down to a LOQ (Limit of Quantification) of 10 nanograms per 25 mm cassette filter and a limit of detection of 2 nanograms per 25 mm cassette filter. As currently produced, lactose also has a dustiness index similar to many high-potency compounds (Note: See Chapter 4 regarding the "dustiness index" research work and surrogate powder results.)

E. Analysis

This section describes the three primary analytical methods used to determine specific chemical compound(s) being measured by particle counting. These analytical methods include chemical speciation, aerodynamic particle size, and optical particle size.

1. Chemical Speciation

As portrayed in Table 2, chemical speciation can be presented either as a specific analyzable chemical entity, as a functional grouping of entities, or as a general class of entities depending on the assay.

In the 1970s the analytical chemistry division of the American Chemical Society (ACS) commissioned a peer review committee to define LOQ and Limit of Detection (LOD) parameters. Based on the best available practices and comparative testing for statistical significance, the following guidelines were established:

LOQ is when the signal to instrument noise meets or exceeds 10:1 ratio
LOD is when the signal to instrument noise meets or exceeds 2:1 ratio

Table 2 Chemical Speciation Techniques

Descriptor	Assay method
Chemical-specific	Mass spectrometry (the preferred method for best results)
	Liquid chromatography (the most sensitive)
	Gas chromatography (less sensitive)
Functional groups	Infrared spectroscopy
	Ion chromatography
General class	Gross chemical assays

Due to constraints imposed by environmental regulation, many laboratories have had to adopt a lower LOQ of 3:1 signal to noise. Based on the ACS peer committee's findings, at a level of 3:1 the instrument signal is not describable with any degree of statistical credibility. However, many trace laboratories use this environmental descriptor of LOQ whenever they perform work in the environmental or occupational health arenas. It would be wise to discuss with your laboratory which threshold they use for LOQ and if they would be willing to provide an LOD parameter.

2. Aerodynamic Particle-Size Speciation

In the absence of significant side-by-side correlation of particle size and chemical analysis data, there is little reason to suspect that they are comparable. The aerodynamic particle size is based on many years of experience in the environmental field using impaction collection devices, which have been calibrated by the American Society of Mechanical Engineers (ASME) procedures. Their reliability and performance have been the subject of many articles. While they are very robust in performance, the data acquired by each different type of instrument is subject to its own specific errors. NIST standard source materials are used for comparative performance testing of impaction devices, and indeed several NIST materials are prepared in size-selective ranges using such devices.

3. Optical Particle-Size Speciation

Selecting an appropriate optical particle-counting method depends on the physical characteristics of the particle under investigation relative to precisely manufactured spheres of polystyrene that are used for calibration purposes. The spheres are of NIST-traceable particle sizes with a selection of narrow particle-size ranges. The instrument performance is calibrated using these spheres as source materials. Consequently, the physical properties of the materials, namely surface reflectivity, particle size, particle shape, and surface characteristics, will all affect how the instrument perceives the particle since the monitoring method measures light scattering from a surface as a single event. An added complexity is the fact that small particles can be hidden by large particles (coincidence phenomenon) or the light from both small and large particles traveling parallel will be summed together as a single incident of larger apparent particle size than either of the original particles.

IV. TYPES OF SAMPLING INSTRUMENTS

This section describes a variety of instruments traditionally used in both occupational health and environmental areas, but which have advantages for containment monitoring purposes. The major categories of instruments to be discussed include gross particle measurement, aerodynamic particle size, optical particle size, isokinetic sampling techniques, fibrous aerosol sensing, and viable organism collection devices.

A. Gross Particle Measurement

Instruments in the gross particle category, described in detail in Table 3, are designed to measure the mass of particles.

B. Aerodynamic Particle Size

Table 4 describes instruments designed to measure the aerodynamic particle size of material. This is primarily accomplished by assessing the impaction of the particle on a known surface.

C. Optical Particle Size

Optical particle-size sampling instruments, as described in Table 5, measure particle size of the unknown material in reference to known calibrated particle-size standards.

D. Isokinetic Sampling Techniques

Isokinetic sampling instruments (see Table 6) attempt to measure particle size based on matched sample velocity.

E. Fibrous Aerosol Sensing

A separate class of instruments, identified in Table 7, has been developed to measure fibrous aerosols.

F. Viable Organism Collection Devices

Viable organism sampling instruments, shown in Table 8, attempt to use growth cultures to measure viable particles.

V. SAMPLE METHOD SELECTION

In deciding on the most appropriate particle-sampling method, a number of factors must be taken into consideration, including particle characteristics, available sampling devices, and sample flow.

Table 3 Gross Particle Sampling Instruments

Instrument	Description	Analysis/operating parameters
Electrostatic precipitator	This gross particle measuring system relies on electrodeposition of particles as they traverse through an electric field within the instrument. This is an old technique that has been out of favor since the 1970s due to serious operational and reproducibility considerations.	Analysis requires collection of the samples from the precipitator. Sampling can be operated in isokinetic mode, but such practices were not recognized at the time the device was in common usage.
Impinger	A true plate and jet impinger relies upon impaction of particles on the plate in a liquid medium. When operating correctly, the liquid in the impinger is displaced by the swirl of micronized air bubbles, the total volume of the air + liquid being at least three times the original volume of the liquid alone. Many people have never seen the plate and jet impinger function properly. The efficiency varies significantly, depending on whether it is used as a true impinger or or as a bubbler. This trend began with the National Institute for Occupational Safety and Health (NIOSH) standards completion program and has become ingrained in industrial hygiene practices since that time. Errors due to the evaporative loss of solution during sampling results in poor sampling efficiency. For effective use the impinger requires constant attention during the brief sampling period (~15 min.).	Analysis may be conducted by filtration of the solid from the liquid, drying, and weighing, and in some cases, evaporation of the liquid to measure the soluble materials present. Chemical analysis can be performed on both of these categories of material. It can be operated in isokinetic mode, but typically is not.

Table 3 Continued

Instrument	Description	Analysis/operating parameters
Cassette	37 mm air-sampling cassettes, both two- and three-section holders, have become a recognized method for most dust-monitoring measurements. The method relies on deposition of the dust on the surface of a filter as air flows through the membranes or micro-orifices of the filter, depending on the design and style of the filter. The particles adhere to the surface of the filter due to both impaction and static charge. Particles bounce off the surface of the filter as a result of entry velocity, and also due to charge repulsion between particle and filter surface result in frequently observed errors—particles being attracted to the cassette casing instead of the filter due to the cassette also being charged. The cassette walls are traditionally not washed down as part of the sample workup procedure. Several investigators have reported errors due to static attraction by the cassette. This error becomes more significant with small particles and potentially is an extreme error in the case of small particles at low-exposure guidelines (e.g., less than 1 $\mu g/m^3$). Other modifications to the method include the use of 25 mm cassette holders and, especially for fibrous materials, the use of extended (2-inch) conductive cassette cowls to mitigate the electrostatic error problems. The 25 mm cassette has been extensively investigated for use in the asbestos industry. Major sampling issues with the cassette are associated with the high differential pressure created across the membrane during airflow. This is typically not an issue for membrane type filters, but micro-orifice filters (e.g., polycarbonate and PTFE membranes) experience this problem significantly. The outcome is a lowered maximum airflow rate and a shorter pump battery life. The design of the filter support system in the cassette base can also have a significant effect on particle deposition, penetration, and bypass leakage.

A 47 mm cassette area is used in a high-flow mode to obtain more samples as the airborne concentrations decrease. 37 mm and 25 mm cassettes are also used in the same high-flow modes with significant constraints. | Cassettes can be operated in isokinetic mode, but typically are not. See IOM in the following text. |
| Piezo balance | A vibrating quartz transducer that collects particles by impaction. This results in a change of frequency of the transducer, which is expressed as a weight equivalent. This was in vogue during the late 1970s, and is no longer widely available. | Relies on particle impaction using a plate and jet approach similar to the impinger. Chemical analysis not feasible. Isokinetic sampling not feasible. Multiple discrete measurements can be made before the sampling medium requires cleaning. |

GCA RDM-101	A paraffin-coated Mylar membrane upon which particles are impacted and measured by beta energy attenuation and referenced to typical standard materials such as graphite. This was in vogue during the late 1970s, and is no longer widely available.	Relies on particle impaction using a plate and jet approach similar to the impinger. Chemical analysis is not feasible. Isokinetic sampling not feasible. Multiple discrete measurements can be made before the sampling medium requires changing.
GCA RDM-102	Relies on collection of particles on a filter and measurement of the change of beta energy attenuation as a result of the presence of the sample. Also referred back to an equivalent beta absorption standard. This was in vogue during the late 1970s, and is no longer widely available.	Uses a disposable filter medium for collection. Chemical analysis feasible. Single sample only. Not isokinetic by design.
Settling plates	A passive (nonisokinetic) monitoring system for collection of room contaminants on a standardized surface area. Somewhat equivalent to swab testing. Can be described as particles per square inch or milligrams per square inch of surface deposition.	Typically uses a 60 mm diameter glass petri dish. Comparable in area with a 4 sq. in. swab test. However, the glass petri dish method has a much higher recovery rate (90–95%) by virtue of the glass surface and solvent washing, approximating quantitative recovery. Swab testing, normally associated with finished metal collection surfaces, typically yields much lower recovery (recovery factors can be as low as 7–10% in some cases, depending on the metal surface finishes). An additional disadvantage of swab testing metal collection surfaces is the potential for mistakenly including legacy data from previous sampling events. Glass petri dishes offer pristine surfaces with each test.

Table 4 Aerodynamic Particle-Size Sampling Instruments

Instrument	Description	Analysis/operating parameters
Andersen impactor	A single-orifice impactor used for environmental monitoring in the below 3 μm range.	Chemical analysis can be accomplished by removal of the sample from the collection disk placed beneath each single orifice. Must be used with a constant flow device. Can be run in isokinetic model.
Sierra Cascade impactor—as used on personal sample pumps	Fixed five- or eight-stage multi-orifice aerodynamic particle-size sampling subject to variable size range based on efficacy of sampling pumps.	Can be run in isokinetic mode. Chemical or weight analysis. Precision and accuracy of particle-size cut point seriously affected by pump. (See Figure 2)
Graseby-Andersen sampler	An eight-stage multi-orifice sampler relying on collection on filter medium or aluminum disks placed below the multi-orifice plate.	Operates at constant flow rate using high-vacuum pump and needle valve or limiting orifice flow control. Isokinetic sampler. Chemical and/or weight analysis.
Cyclone	Device for precutting the larger particles from a sample flow to enhance the small particle size sensitivity and specificity of a sampling system.	Can be run in isokinetic mode. Both large particles (from grit pot) and small particles can be chemically and/or weight analyzed.
TSI Piezo-balance with a cyclone	Cyclone is used to cut out particles above a selected particle size range, resulting in a more narrowly defined particle-size range on the sensor.	Plate and jet impingement sample not recoverable. Not designed for isokinetic operation.
GCA RDM-101 with a cyclone	Cyclone is used to cut out particles above a selected particle-size range, resulting in a more narrowly defined particle-size range on the sensor.	Plate and jet impingement sample not recoverable. Not designed for isokinetic operation.
GCA RDM-102 with a cyclone	Cyclone is used to cut out particles above a selected particle size range, resulting in a more narrowly defined particle size range on the sensor.	Sample recoverable for chemical analysis. Not designed for isokinetic operation.
Plate and jet impingers	The particle size segregation is determined by the distance of placement between the jet and the plate. While commercial devices are not available to achieve this, it is possible to change the sampling parameters using custom glass-blown devices. Not in common use.	Can be run in isokinetic model. Chemical and weight analysis feasible.
Vertical elutriator	A device relying on air stratification and particle-settling rate (e.g., cotton dust) to provide information on the more gross aspects of airborne behavior. It is not really true particle size; it relies on settling rate for sample collection.	Visual identification.

Table 5 Optical Particle Size Sampling Instruments

Instrument	Description	Analysis/operating parameters
MIE (GCA) particle counter (RAM 1) with or without a cyclone	The particle counter presents data by converting light flashes as counted by the instrument to an equivalent mass of NIST-traceable Arizona road dust. Use of the cyclone allows the instrument to be selective to the sub-3 μm particle range (i.e., respirable equivalent particle-size range).	Particle reflection of a laser beam. Scrolling sample data, with adjustable integration periods.
DATA RAM with or without a cyclone	The particle counter presents data by converting light flashes as counted by the instrument to an equivalent mass of NIST-traceable Arizona road dust. Use of the cyclone allows the instrument to be selective to the sub-3 μm particle range (i.e., respirable equivalent particle-size range).	Particle reflection of a laser beam. Scrolling sample data, with adjustable integration periods. Data logging and user-friendly on-board data reduction.
HAM (handheld aerosol monitor)	This is a passive optical particle-counting device, which relies on the ambient airflow to transport particles through the sensing area.	Relies on manual observation of a continually changing data readout.
Grimm	The current industrial hygiene version of the Grimm presents fixed particle-size ranges using a 1.2-liter per minute sampling volume with sample entrance through two parallel plates above the instrument.	Nonisokinetic. User-friendly on-board data presentation.
CLiMET	Fixed particle-size range instruments typically using a 37 mm entrance probe at one ACFM flow rate, equivalent to a 90 linear feet per minute air velocity. The CLiMET can also be purchased with different airflow volumes. 2-, 6-, 16-, and 18-particle size channel devices available.	Designed to be run in isokinetic mode. Limited data screen. Has a computer interface for data output. Unique optics (patented) provide high sensitivity.
Visual	Empirical human gross characterization based on like scattering (<3–5 μg diameter) and like blocking (>25–100 μg diameter) behavior of particles.	

Table 6 Isokinetic Sampling Instruments

Instrument	Description	Analysis/operating parameters
IOM	A device developed at the Institute of Occupational Medicine in Edinburgh, Scotland that attempts to match the sampler entrance velocity with a typical air velocity in the work zone. Although it is subject to the performance of the battery-operated sample pump, nonetheless it is an improvement over the typical cassette, yielding data 2–3 times higher than a typical closed-face cassette for a typical work environment. Available in either a single-hole or nine-hole configuration. Subject of significant research effort. This is a UK regulatory tool for workplace compliance measurements.	Chemical and weight analysis feasible.

Table 7 Fibrous Aerosol Sampling Instruments

Instrument	Description	Analysis/operating parameters
MIE fibrous aerosol monitor (FAM-3700)	An optical counting device specifically designed for measuring fibrous aerosols. The detection relies on presenting the faces of a fibrous crystal to a light beam with detection at an off-angle. This is achieved using a rotating magnetic field to spin the fiber during transit past the light field.	Selectable aspect ratio for a particle, filament, or aerofoil.

Table 8 Viable Organism Sampling Instruments

Instrument	Description	Analysis/operating parameters
Spinning disk viable	A nonisokinetic gross-sampling system typically used in hospital and clinical environments.	Incubation, inoculation, and microscopic examination.
Graseby-Andersen viable sampler	A five-stage particle impactor tuned to provide separation between the three common airborne organism size ranges. This affords greater discrimination of organisms during culture and growth.	Typically isokinetic sampling. Inoculation, incubation, and microscopic examination.
Settling plates	Means of transferring to culture media for identification purposes.	Passive sampling on either bare or media-coated plates.
Sticky tape	Used either as a settling surface (see Settling plates) or by contacting with a surface to collect organisms (swab equivalent), which are then separated for culture and/or identification.	Inoculation, incubation, and microscopic examination.
Liquid impactor	A jet impinger where the impingement surface is the surface of the liquid collection medium to mitigate cellular damage.	Inoculation, incubation, and microscopic examination.

Note: Applications of the devices in Table 8 have been further extended to enhance accuracy of observations. For example, introducing a sample through a slit onto a moving substrate, yielding a time-dependent data set.

A. Characteristics of Particles

Particles can be characterized in a number of ways:

Chemical identity
Particle size
Particle shape
Microscopy–discrimination approaches based on illumination source selection and media preparation

Particle-size distribution
Liquid aerosols
Solid aerosols

Chapter 4 provides a more in-depth discussion of particle characteristics and their impact on dustiness.

B. Devices

This section highlights features of the different sampling devices available.

1. Personal Sampling Pumps

Battery operated. Typical metastable flow rates 1.7–4.0 L/min. Sampling duration ranges from minutes to hours. ("Metastable" here refers to the effect on flow rates from such long-term variables as temperature, humidity, filter loading, and battery power.) To the extent that these variables are "stable" within a given period of time, the pump's flow rate is constant. However, as any of these parameters shift, for example over several days of monitoring, the pump's flow rate might also shift from previous days, even though it is observed to be "stable" during a multiple-hour sampling period during a single day.

Low-flow AC operated. Typical constant flow rates 1.7–4.0 L/min. Sampling duration can extend to days.

2. High-Flow Pumps

Battery operated. Unstable sample drawing devices for short-term sampling. Typical flow ranges of 12–18 L/min. for up to 30 minutes.

Line operated. More stable, but still metastable, flow rates in the 12–18 L/min. range for periods up to an hour.

3. Flow-Control Devices

Restricting orifices: working subsonically, are subject to small pressure-differential effects.

Limiting orifices: working at sonic velocities and stable as long as gross pressure effects do not occur.

4. Sampling Manifolds

A sampling manifold (described in detail in the Specialized Monitoring Techniques section) distributes airflow through a series of noncollapsible tubes and connectors from the sample point to a vacuum source. Flow control is achieved using either a flow restrictor or a critical flow device to establish a constant and reproducible sample flow rate, subject to constraints by the laws of physics (fluid dynamics, specifically). Actual flow is differential pressure limited or related.

C. Sample Flow Considerations

Sample flow considerations are an outcome of the sampling scope and strategy as described in the Sampling Strategy Design and Execution section. Major decisions are required to determine appropriate sample flow. Isokinetic sampling is a major consideration. Particle monitoring is not a precise science. It is significantly affected by airflow conditions external to the sampling device. Consequently, the resulting data are a function of the sampling flow rate and the standards used to interpret exposure or contamination levels. Major research led to the development of the Institute for Occupational Medicine (IOM) sampler to mitigate errors due to non-isokinetic sampling. Independent studies also have shown significant differences between isokinetic and nonisokinetic sampling strategies in workplace environments. A pseudo-isokinetic sampling cassette can be made by calculating an appropriate opening size for the cap of a cassette to establish matching air velocities at a given sampling volume (e.g., 18 mm diameter opening for a 10 L/min. flow rate for 90 linear ft./min. simulation). The cap of the cassette can be drilled or bored to this opening and used to approximately replicate the data obtained from an IOM sampler. Refer to the Calibration of Air Sampling Devices section for appropriate calibration strategies.

All sampling devices are subject to pressure changes—some gross, some small—and no system should be operated with the assumption that it will remain stable unless pump calibration or in-line pressure monitoring is used to measure the performance throughout the entire sampling period. Pressure fluctuations may result from instability of the sampling media, from sampling media pluggage during sampling, or from nonrobust connections within the sampling train. Calibration of any air-sampling system should be performed with open tubing between the calibration device and the sampling system such that a minimal pressure drop is introduced during calibration. See earlier discussions about soap-film flowmeters, rotameters, etc. in the Calibration of Air Sampling Devices section.

Total sample volume collected must relate directly to the precision and accuracy of the analysis system and the LOQ or the LOD depending on the data needed. These are measurable quantities in the case of chemical speciation and gravimetric analysis. While they are not precisely measurable for OPCs, particle counters are usually calibrated for a single flow rate and should be used as such for any reference calibration to be valid (e.g., polystyrene beads calibration).

VI. SAMPLING STRATEGY DESIGN AND EXECUTION

A. Sampling Scope and Definition

It is critical to be precise about the scope and definition of the sampling end result. This will lead to an appropriate sampling strategy based on relative strengths and weaknesses of different sampling approaches and devices. Considerations to be taken into account when defining the scope:

> Anticipated exposure or contaminant level
> Contaminant or exposure guideline
> Nature of the particulate (e.g., density, optical reflectivity, stability, surface charge, particle-size distribution, particle shape, and equivalency to reference standards)
> Time-weighted average data/sample integration period (i.e., what is appropriate to address the experimental question at hand?)
> Operation time, cycle, and frequency of transients
> Isokinetic issues if the particle is aerodynamically greater than 5 microns in diameter
> Personal sampling
> Area sampling
> Gross contamination build-up with time
> Does it really replicate the workplace?
> Standard sampling practices or equivalent
> Bulk particle size versus airborne particle size—expect significant differences
> True particle density
> Background contamination due to the facility rather than the active compound (i.e., what is the activity to be sampled? Does the activity generate its own dust, other than active compound?)

B. Sampling Strategies

Case studies, presented in Chapter 6, illustrate a variety of sampling strategies for different challenge environments that have been used in actual practice:

> Comparative hood study
> Parallel sampling
> Fog generation
> Outdoor facility, no analytical assay
> Containment capability assessment (evolving, based on health physics principles)

Parallel data-cassette samples and OPCs
OJT (on-the-job training) and performance testing
Process-generated liquid-aerosol sampling

C. Data Interpretation

Figure 3 shows an example of interpreted data output from the six-channel CliMET OPC. The left-most axis shows particle counts that have been converted into an average mass value (in $\mu g/m^3$) per minute. The right-most axis displays the color-coded particle-size ranges (in microns), which correspond to the legend, appearing below the chart. The bottom axis displays the time. This graph was created using the mean particle volume for each size band, based on particle volume being $4/3\pi r^3$, ''r'' being particle radius. The graph is normalized to a particle density of 1.0 g/cc. Actual comparison data should be generated using a known reference sampling system, bearing in mind all the eqivalency considerations such as isokineticity, large versus small particulate contributions, etc.

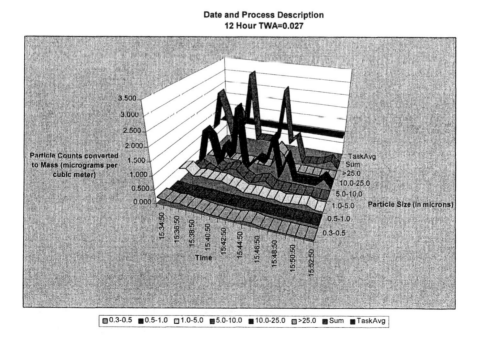

Figure 3   Data from an optical particle counter.

VII. SPECIALIZED MONITORING TECHNIQUES

A. Sampling Manifold

1. Scope

To use a common vacuum source to draw airflow from multiple sample locations in a controlled manner.

2. Advantages to a Manifold System

1. Expand sampling capability without additional pumps.
2. Sampling is not compromised by battery run life. (The assumption here is high-volume pump with AC power.)
3. Simultaneous start/stop times for all sample points.
4. Less flow fluctuations than a traditional sampling pump.
5. Different stable flow rates available at each sampling location along a manifold.
6. Single electrical outlet for multiple samples.
7. Higher level of sample integrity through constant, nonpulsating flow.
8. More reproducible analyte recovery in the laboratory.
9. Any common system variability is directly proportional for every sampling device along a manifold.
10. System alterations are traceable, repeatable, and recalibratable as long as good written records are retained.

3. Principle

The basic concept is to distribute airflow through a series of noncollapsible tubes and connectors from the sample point to the vacuum source. Flow control is achieved using either a flow restrictor or a critical flow device to establish a constant and reproducible sample flow rate. Actual flow is differential pressure limited or related. In the limited flow condition, there is a natural stable limitation when the linear airflow through the critical flow device attains sonic velocity. Flow restrictors attain a reproducible stable flow when the specific differential pressure is repeatedly established across the restrictor.

4. Materials Needed

Vacuum Source, Typically:

1. A high-flow pump, battery or AC operated, having low-suction pressure capability (e.g., a Gillian high-flow pump). Note, this vacuum source should be used for short operating times only, and where a

shallow suction pressure is required, down to -16 in. water pressure.

2. A carbon vane rotary vane pump having high-volume capacity and high negative-pressure capability (e.g., a Gast™ rotary vane pump delivering an airflow of 35 actual cubic feet/minute (ACFM) at barometric pressure but capable of operating at -26 in. of mercury pressure.

3. A house vacuum system, such as that used to drive process equipment, held at constant header pressure.

Manifold, Typically Made of:

1. Metal tubing and fittings, such as copper, galvanized or stainless steel
2. Rigid PVC pipe and fittings
3. Reinforced vacuum hose (kink-proof) and fittings
4. A hybrid of any of the previously mentioned

The manifold distributive system has to be such that there is adequate negative pressure distribution to all flow control devices at each sample location, wherever they are, without stalling the pump.

Negative-Pressure Measuring Device, which Can Be:

1. Oil or water manometer for negative pressures in the range of 3 ft. of water
2. A gauge capable of reading in inches of water (for Gillian) or inches of mercury vacuum, negative atmospheres, or Torr (for house vacuum or pump)

Sample Flow-Limiting Devices that Either Can be Purchased Commercially or Fabricated on Site, Typically Made of:

Restricting flow limiters—copper tubing packed with glass wool for use with low capacity pumps

Critical flow limiters—based on sonic flow (speed of sound) being consistently achieved in a precision bore orifice or tube

1. Commercially made flow-limiting orifices (e.g., Millipore Corp.)
2. Homemade orifices consisting of metal tubing filled with cast epoxy adhesive and drilled through its length with a drill bit or smaller reamer
3. Homemade orifices consisting of standard-gauge hypodermic syringe needles broken off from the attaching point to the syringe and bonded into a piece of rigid tubing with epoxy adhesive
4. Precision ruby orifices

Sampling Device Adapter, such as:
1. Commercial cassette adapters with threads for holding limiting orifices (e.g., Millipore Corp.)
2. Standard Luer #2 taper cassette adapters made of steel (e.g., SKC)
3. Rubber vacuum hose of 1/4 in. internal diameter (i.d.) to stretch over the cassette adapter
4. Thick-wall vacuum-rated Tygon tubing of 1/4 in. i.d.

Nylon cassette adapters are not recommended because of their compressibility and undesirable effect on pressure drop.

Sampling Device, such as:

1. 24 mm cassette with standard Luer #2 taper fittings
2. 37 mm cassette with standard Luer #2 taper fittings
3. 24 mm or 37 mm cassette used open face
4. IOM sampler for isokinetic sampling
5. Particle-size sampler

Sample Support, such as:

1. Camera or lighting tripod
2. Wall hanger
3. Roof hanger
4. Adhesive tape
5. Tie wrap
6. Table

Pressure Controller: An *in-line* vacuum relief for a flow-compromised pump.

Flow Calibrator: A low Δp calibrator (<1/2 in. water) designed for the selected flow rate with a low Δp connector attachment of the cassette. Tubing for Luer #2 fittings and sealable connector for all others.

5. Procedure

1. Lay out the manifold according to the sampling plan.
2. As shown in Figure 4, ensure that the manifold header at the vacuum source has the largest bore of manifold delivery tubing, only reducing to the smaller sizes, where it is delivered to each unique sampling location. From practical experience, a header interior diameter of 0.75 in, should be considered as a minumum.
3. Connect the hose and fittings beginning at the vacuum source, working out towards the sample locations.

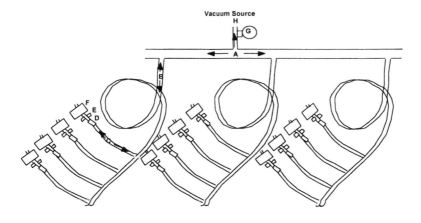

Code	Description	Parameters
A	Largest Bore Tubing	¾" i.d.
B	Medium Bore Tubing	¾" or ½" i.d.
C	Smallest Bore Tubing	¾", ½", or 3/8" i.d.
D	Restrictor, or Flow Limiting Orifice (sonic)	14, 10, 4.9, 3 or 2 lpm commercial or any custom orifice
E	Cassette Adapter	As Necessary
F	Sampling Cassette	As Necessary
G	Vacuum Gage or Manometer	As Appropriate
H	Vacuum Source Inlet	

Figure 4 Sampling manifold for a 35 ACFM rotary vane pump system or house vacuum.

4. Secure the flow-limiting devices and adapters to the manifold at each sample location.
5. Attach representative sample cassettes to each adapter.
6. Operate the pump and ensure that each sample location delivers appropriate airflow using the calibrator. Adjust flow at each sampler if the flow is not sonic at the restrictor.
7. If inadequate performance, troubleshoot for leaks and sufficient vacuum drawn at the head of the pump. Remedy as necessary.
8. If insufficient airflow throughout without system leaks, decrease the number of sample locations or increase the capacity of the vacuum source (e.g., change to a higher flow-range pump, or split the locations using additional pumps, or connect additional pumps in parallel onto the same manifold).

6. Manifold Options

There is more than one option for applying the principles laid out previously for sampling manifolds. Two options are presented here.

First Option: The Limited Negative Pressure Application.   Due to the development of vacuum in the inches-of-water range, it is not possible to use critical orifices or similar devices that rely on pressure drops of the order of 10 or more inches of mercury (>5 psi). Matching flow-restricting devices are used in this application. The simplest of these are pieces of rigid tubing (e.g., soft copper, packed with class C [laboratory grade] glass wool of nominal 8 μm diameter). Matching of flow restrictors is made using a simple Slack Tube Manometer (a piece of polyethylene tubing with water in it and a measurement scale) as shown in Figure 5.

A self-adjusting sample pump is set at the selected flow rate for the sampling device (e.g., 2 L/min), and the flow restrictor is connected to the inlet to the pump through a tee also connected to the measuring device (e.g., a manometer in the drawing). The flow restrictor is packed tightly with glass wool using a tamping device (e.g., a wooden or glass rod), until a pressure drop of 10–12 inches of water is achieved. This value is determined by the operating pressure range of the high-flow pump being used for the manifold and should typically

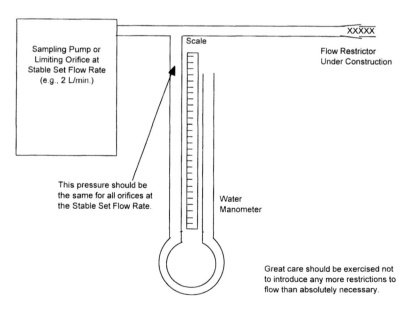

Figure 5 Preparation of flow restrictors.

be set at a value of at least 1/3–2/3 of the operating pressure of this pump. (Typical values for a Gillian high-flow pump operating at −16 in. of water is −10 in. of water across the restricting device.) All restricting devices to be used in the system are constructed in the same way to achieve the same pressure drop at the same flow rate.

It is feasible to use different flow rate restrictors in a single manifold. The differential pressure across each restrictor at its ''selected'' flow rate must be within ±0.1 in. water pressure of all other restrictors connected to the same manifold. This requires fabrication of the device using the ''selected'' flow rate and using the same setup. This assumes the manifold pressure is constant at every tubing takeoff.

The restricting devices are attached to the sampling manifold at the designated sample points. All sampling devices are attached to the inlet of the appropriate restricting devices, and the flow through each sampling device is calibrated using an appropriate calibration system of extremely low pressure drop, such as a soap-film flowmeter (bubble tube). All sampling devices have to be attached before a single device is calibrated because the system as constructed is ballistic in nature. Any small fluctuation in pressure within the manifold due to a flow-restricting device or unusual sampling device pressure-drop change will affect all other flow rates of the sampling devices in a proportional manner. During use, the status of the high-flow pump, manifold, and sampling array is monitored by the manometer or gage placed adjacent to the pump inlet (H) in Figure 4. A change of more than 0.3 in. water pressure is cause for concern, and requires recalibration of each sampling device to ensure validity.

Using low suction pressure sampling pumps requires them to be plugged into an AC power supply because of the variable flow resulting from the use of battery-operated equipment.

The purpose of the restrictor is to

1. increase the sampling flexibility of the pump or vacuum source, allowing the flow rate to be reduced without loss of performance;
2. dampen flow fluctuations at the cassette air inlet;
3. reduce flow fluctuations for the entire manifold assembly; and
4. allow for a single vacuum source to control (start, stop) multiple samples for data and recordkeeping consistency.

Any flow change at the head of the pump will be reflected in a relative change at every sample point. This is due to the change in vacuum head at the restrictor outlet to the pump, achieving a directly proportional flow rate change at each sampler head and maintaining qualitative equivalency even when change occurs in the sample system.

In use, the sampling devices are attached to the manifold and are precalibrated. At the time of starting the operation or sampling schedule, the sampling pump is turned on, initiating sampling at all sampling device inlets. At the conclusion of sampling, stopping the pump terminates all sample flow to all sampling devices.

As a point of time conservation, it may be practical to postcalibrate each of the sampling devices before the pump is turned off. Alternatively, the pump must be restarted after resting at the end of the sampling period. The compromise of this alternative would be any disturbance of the sampling device heads (i.e., the device and the restrictor, due to flow interruptions or battery power variations).

Second Option: Full-vacuum Manifold System. The configuration of the manifold follows the same pattern as the previous example with the following critical differences:

The pump is a high volume vacuum pump, typically of the rotary vane type.
A pressure regulator to indicate the vacuum at the intake of the pump.
There is an adjustable pressure relief valve to extend the range of the pump without additional wear.
An orifice or similar critical flow device is located at each sampling head.

The principle of the system relies on achieving critical flow at each of the orifice/critical-flow devices. This typically requires approximately 10 psi of vacuum at the inlet of the pump. Caution must be taken when using sampling filters of the micro-orifice type, such as teflon and polycarbonate filters, which behave like flow-limiting orifices and do not allow the critical flow device to attain criticality. The vacuum at the inlet pump is maintained throughout sampling at a level greater than that required by the critical-flow devices, thereby ensuring that the sample flow will be constant and reproducible every time the manifold is in operation. Calibration consists of pre- and postcalibration of the entire manifold during the first run and observation of the vacuum at the inlet to the pump during each subsequent sampling survey.

7. Recommendation

If at all possible, resort to using a full-vacuum manifold system. System stability typically will be better established using critical sonic-flow devices rather than restricting flow devices.

B. Aggressive Air Monitoring

Aggressive air monitoring is adapted from the Asbestos Hazard Emergency Response Act (AHERA) regulation adopted as a means of ensuring effective cleanup of an area after an asbestos-decontamination event. It was adapted for the pharmaceutical industry after it was found that conventional swab sampling on different surfaces (e.g., glass; 2B mill finish stainless steel, and 20, 40, 60, 100, 120, and 180 grit stainless-steel finishes) was unrepresentative of the residual contamination left on them. This information was obtained using a controlled series of statistically designed surface inoculation and swabbing studies.

The principle is to clean a facility or suite and allow it to dry completely. Then install a statistically random set of samples inside the ''clean'' suite and a comparable set of random samples on the supply air side of the suite. Sampling is started and high-pressure air (90 psi) is blasted onto all surfaces of the suite from a distance of 3 in. to disturb any adhering surface contamination from the walls, ceilings, floors, equipment, etc. Sampling is continued for approximately two hours after the air blasting is completed to ensure continuous sampling during concentration decay of any airborne contaminants in the room. After analysis, the statistical data sets for the suite and the supply air side of the room are compared to see if there is a statistically measurable difference between the two. If the contamination level in the room is ''statistically'' higher than the supply air side, the room is deemed contaminated. When the supply air side is statistically higher than the room (an event not uncommon in a typical building), Quality Control needs to make a judgment on whether this is a gross facility contamination in spite of the containment suite being clean. The method is statistically robost when using five sample locations in both the containment suite and the supply air side, providing an 85% confidence level to the data.

If the suite is contaminated with a pharmaceutical active, it is prudent to wear full PPE during the aggressive monitoring, with appropriate decontamination of the suit and operator after the aggressive monitoring event. PPE should also be worn while setting out the samples initially and recovering them after the event is completed.

6

Particle Monitoring: Case Studies

Brian G. Ward

Eli Lilly and Company, Indianapolis, Indiana

This chapter demonstrates a progressive learning curve that has occurred over the years with regard to particle monitoring, in several of its various forms, and provides case studies demonstrating how particle monitoring has been used in real-world applications. Below is a list of the case studies and key particle-monitoring advancements discussed in this chapter.

Liquid Aerosol Issue Involving Vegetable Oil (ca. 1975): Aerodynamic particle sizing, Impactor sampling

Coal Tar Pitch Volatile (CTPV) Fugitive Emission (ca. 1979): Sampling with impactors of different diameters

Air Clearance Monitoring after Decontamination (ca. 1992): Aggressive air monitoring

Dispensing Operations (ca. 1993): Real-time optical particle counters (OPCs)

New Facility Commissioning (ca. 1993): Combination settling plates, air sampling, and ventilation flow

Performance Qualification of a Seeding Isolator (ca. 1994): OPC leads to operator buy-in

Rigid Isolator (Glovebox) Performance Testing (ca. 1995): Adapted AH-ERA sampling protocol

Outdoor Processing (ca. 1995): OPC provides performance feedback

Operational Qualification of a Rigid Isolator (ca. 1996): Validation with OPC of empirical and quantitative testing

Full Facility Commissioning during Extended Production (ca. 1997): Full-time OPC monitoring

Particle Control Test Comparison (ca. 1997): Settling plates used to evaluate cross-contamination

PPE Performance Testing (ca. 1997): Extreme sensitivity of OPC

Contained Packaging System—Initial Development (ca. 1997): OPC facilitates rapid prototype system development

Please note, drawing direct comparisons between different data is inappropriate
since the devices used to collect the data, as well as the context in which they
were used, are different.

I. LIQUID AEROSOL ISSUE INVOLVING VEGETABLE OIL (ca. 1975)

A. Background and Scope

This application was in a workplace where nylon fiber was being spun, and a
workplace health-related problem was identified. This situation could arise in
any facility with an HVAC system and that handles or prepares solutions, oils,
suspensions, and/or emulsions (e.g., a private residence, a hotel, a pharmaceutical
plant). The situation was presented as four cases of lymphocytic interstitial pneu-
monitis. The scope of the study was to identify the causative agent. Although
the work environment was not a pharmaceutical facility, this example illustrates
how particle monitoring can be used to identify the causes of health problems,
which can occur in any industrial setting.

B. Method

Initial sampling was done using conventional mixed cellulose ester filter membranes in conventional 37 mm diameter cassettes, followed by chemical-specific
analysis of vegetable oil mist. A first pass (nonexhaustive) sampling approach
to rapidly characterize the causative agents was adopted. Primarily, a photo-ion-
ization detector (PID*) was used to monitor for hydrocarbons, and an RDM 101
OPC and a Piezo balance were used to monitor for particles. The data from these
devices showed no hydrocarbon vapors above the detection limit of the PID and
particle monitors showed concentrations of the same order as the cassette data.
However, neither the cassette, the RDM 101, nor the Piezo balance were sampling
in an isokinetic manner. Isokinetic sampling was not a standard workplace practice at the time. Because of the prevalence of a visible fog in the workplace,
more discriminate sampling was performed using a Graseby-Andersen cascade

---

* PID instrument uses an ultraviolet lamp to cause electron loss from a molecule which in turn is
detected as a change in charge of the airflow using an electrometer.

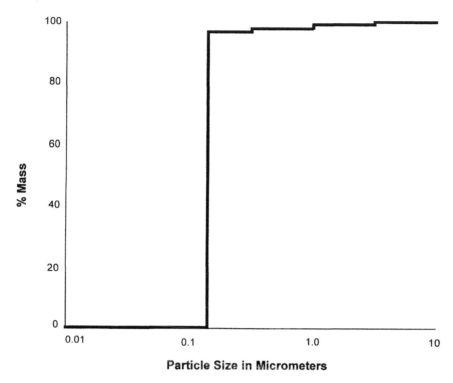

Figure 1 Percent mass by particle size.

sampler consisting of eight particle size stages and a final HEPA filter. This device was sampling at 1 actual cubic foot per minute (ACFM) in a pseudo-isokinetic* mode.

C. Results

Initial sampling with conventional cassettes showed the presence of 1.0–1.5 mg/ m³ of vegetable oil. Based on several samples in the workplace using the Graseby-Andersen sampler, chemical analysis showed a collected mass of vegetable oil in the range 5–8 mg/m³ with predominantly 0.2 μm particles (~95% of the mass as depicted in Figure 1) with no measurable deposition on the HEPA filter. This is an atypical result for a multisource system. Because of the failure of the mixed

---

* Pseudo-isokinetic: A generalized attempt to match the entrance velocity to the instrument/device sample tip to the ambient airflow surrounding it.

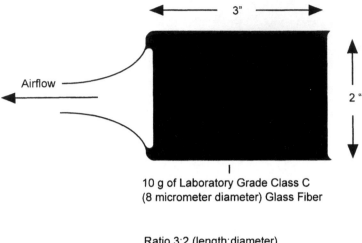

Figure 2 Glass fiber impactor for 1 ACFM airflow.

cellulose ester sample cassette to collect the same aerosol concentration that was later collected on the Graseby-Andersen, it was believed that the HEPA filter in the Andersen sampler might be suffering from the same error. Andersen samples were then drawn using a glass-fiber impactor (Figure 2) in place of the HEPA filter, at which time a measured concentration of 15–20 mg/m³ of vegetable oil was found, with approximately 70% being located on the glass-fiber impactor. Sampling using National Institute for Occupational Safety and Health (NIOSH) and conventional sampling strategies understated the aerosol challenge approximately tenfold based on this finding.

D. Tentative Conclusion

The explanation for this phenomenon was twofold:

    1)   When early health effects were observed and the recommended guideline of no more than 2 mg/m³ of vegetable oil was established, the processing equipment was spinning nylon fiber at a much slower rate than current practices. This resulted in low concentration of aerosol nebulized due to the slow speed of the equipment. As the equipment speed was increased to step up throughput, sheering of oil droplets from the fiber occurred at a much higher frequency and velocity, resulting in smaller and more numerous droplet formation.

    2)   Aerosols are highly stable due to their surface charge (ZETA poten-

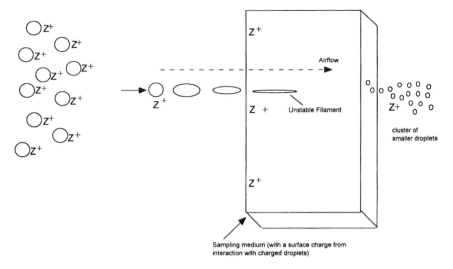

Figure 3   Filament passing through membrane.

tial), which precludes particle collision and growth. As a charged droplet approaches the sampling medium, which is already coated with a surface charge due to droplet collection, it will pass through the membrane as an unstable filament, emerging as a smaller droplet than the original (see Figure 3). For droplets at or below 0.3 μm in diameter, this breakthrough can occur with up to 99% efficiency (in other words, a filter collection efficiency of 1%) and was displayed by this study. This phenomenon is well understood in industrial fume or mist collection, but is not normally of significance in the occupational environment. The glass-fiber impactor replicates the function of the commercially available Brink-type mist eliminator used in industry to quantitatively collect submicrometer mists.

E. Final Conclusion

The ultimate causative agent, in this case, was found to be an airborne concentration of a Legionnaires' disease–like organism, a thermophilic actinomyces, for which the vegetable oil was providing a nutrient base. Innovative particle-sampling procedures provided a new resource and understanding. Benefits of the study included upgraded chest x-rays, lung perfusion, and radioimmunoassay testing to identify any additional cases at risk. Remediation involved cleaning the air-handling system, especially the demister, and changing the type of biocide used in the demister at three-month intervals.

II. COAL TAR PITCH VOLATILE (CTPV) FUGITIVE EMISSION (ca. 1979)

As in the previous case study, although the work environment was not a pharmaceutical operation, this example illustrates how particle monitoring can be used to identify the causes of problems. Such problems can occur in any industrial setting where solutions, oils, suspensions, emulsions, and shock-cooled vapors are involved in the processing step.

A. Background and Scope

Continuing the learning curve of the previous case, examination was done on a CTPV fugitive emission. The emission from a coking operation was a deep-yellow fog that, when sampled with the standard NIOSH procedures, showed a $1-2$ mg/m^3 concentration of CTPVs. The scope of work was to establish whether the NIOSH sampling procedure was effective in measuring CTPV.

B. Method and Results

Replacement of the standard NIOSH procedure with a glass-fiber impactor, developed in the previous study, was utilized. During the study, impactors of different diameters were used during parallel sampling with the conventional cassette. Optimum collection efficiency at a flow rate of 2 L/min. was achieved using a half-inch internal diameter impactor of 3 in. packed length. Measured concentrations exceeded 20 mg/m^3 CTPV, a full order of magnitude greater than previously measured. Details of the impactor are shown in Figure 4. The impactor tube is

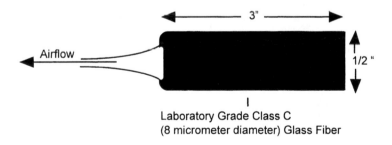

Figure 4   Glass fiber impactor for nominal 2 L/min. airflow.

packed to a density yielding 12 in. of water pressure drop at 2 L/min. flow rate, using Laboratory Grade Class C (0.8 μm diameter) glass fiber.

C. Conclusion/Observation

In both of these cases, had we had OPCs, we would have better quantified and identified the causative factors at the outset. Use of the impactors assisted in creating an effective measure of CTPV releases with subsequent control measures. The actual risk was significantly understated using existing measurement protocols to monitor the deep lung-inhalable challenge.

III. AIR CLEARANCE MONITORING AFTER DECONTAMINATION (ca. 1992)

A. Background and Scope

The scope of this effort was to measure the residual concentration of materials remaining on facility surfaces after decontamination operations were completed. During manufacturing and especially during decontamination and campaign changeovers, occasional release events can be anticipated. Whenever a fugitive release occurs, it is necessary to decontaminate the production area as rapidly as possible after the event is over to mitigate further spread of contaminant. Typical environmental measurements are a factor of 100 times cleaner after crossing a physical barrier, such as a door, into or out of an airlock. In the cases illustrated, there were two, and sometimes three, doors between the hallways and production areas. Consequently, the processing zone can be of the order of 10^6 times cleaner than the hallways. Sensitivity of the instrument is greatly enhanced due to operating in a processing area that is already at the low, or "clean," end of the scale with regard to typical contamination rankings. The sensitivity is what makes this approach feasible.

B. Sampling Strategy

To confirm that an area has been cleaned to the level of shirtsleeve occupancy, a monitoring system that can achieve high sensitivity over short sampling durations is needed. The CLiMET CI-500 is one example of a well-suited instrument for this purpose. The strategy used for monitoring is to take a random sampling of locations in an adjacent clean zone from which the supply air is drawn into the processing suite. Typically, three locations are used to establish the clean zone background, but more could be selected if greater statistical robustness is required. The instrument is then protected (in a large plastic bag) and the protected operator takes the instrument into the processing suite. Sampling is conducted on a predetermined number of statistically randomized locations plus one location adjacent to, but downwind of, the previously contaminated area. If additional statistical confidence is required, a sample set can be taken in an adjacent, but clean, processing zone.

1. Case 1

During early manufacturing of a chemical intermediate (Guideline Exposure on the order of 1 $\mu g/m^3$), a number of fugitive releases had occurred, giving rise to visible surface contamination in a processing room. An initial regime of pressure washing with water was followed by aggressive air monitoring (as described in Chapter 5), with air-sampling cassettes located within the room area and by the air-exhaust intake. After the first room cleaning, passive (ambient, not aggressive sampling) dust levels in excess of 100 times the guideline were measured. Following a second cleaning operation, passive dust levels of 5–10 times the guideline were achieved during aggressive air monitoring. Following a third pressure washing, passive dust levels were reduced to the guideline level. The whole cleaning operation took place over a period of 10 days. An improved cleaning method (ca. 1993) was used in the same facility after a comparable spill. Aggressive sampling returned results below LOQ (Limit of Quantification). In other words, using aggressive sampling and monitoring (reference the Asbestos Hazard Emergency Response Act [AHERA] guidelines) the facility returned to shirtsleeve service after only a 3–day delay.

2. Case 2

The scope was to prove cleanability of a newly designed and built dry-powder dispensing room to handle materials having permissible exposure levels within a 1–100 $\mu g/m^3$ range. After completion of construction, the Quality Control group required proof of cleanability of the suite for an intermediate potency material (i.e., between 1 and 100 $\mu g/m^3$). A sampling regime, using the AHERA guidelines for clearance monitoring, was introduced at this time. Five statistically randomly located control samples were placed in the homogeneous areas (reference the AHERA guidelines) of the gowning room outside the dispensing suite and 5 statistically random samples were located in the homogeneous areas of the dispensing suite. One additional sample and an OPC (a CLiMET CI-500) were located at the HEPA air intake to provide an overall marker for an airborne release event.

A background was established by sampling for 8 hours in the new facility, returning no measurable concentrations of the surrogate material in either the gowning room or the dispensing suite. The walls of the dispensing suite were inoculated with an aqueous suspension containing 10 g of a water insoluble surrogate material using a two-fluid spray gun. Visible staining was observed in several areas of the suite. An operator in protective clothing cleaned the suite using a predetermined and standardized combination of fogging and misting methodology adopted during previous studies. Immediately after the room was dry, a new set of aggressive samples were taken at the previous sampling locations. The investigator donned protective clothing and, using 90 psi air supply delivered through a small jet (in this case a commercial engine degreasing gun), blasted all contaminated surfaces in the room at a distance of 3 in. from the surface. The purpose was to evaluate the maximum potential release of material remaining on the surfaces after cleaning.

When the analyses were returned from the lab, all the cassettes in both control and dispensing zones returned less than quantifiable results—including the HEPA intake. The quantification level was 50 nanograms per cassette, which translated to 0.05 $\mu g/m^3$ for the 8-hour duration of sampling (i.e., 20 times lower than the lowest target level of 1 $\mu g/m^3$). However, the particle counter had shown finite increases of particle counts in the greater than 1 micrometer range while the load cell check plate was being air blasted. Therefore, prudent practice dictated that additional cleaning of the check plate was indicated until no further spontaneous releases occurred even though no measurable concentrations were found on the cassette. The cycle of inoculation, cleaning, and monitoring for the dispensing suite was repeated three times using different department operators. In each case no measurable concentration of surrogate was found. This represented overachievement in that the combination of cleaning procedure and facility design demonstrated higher-potency cleanability than was originally intended in the scope of the facility. At the request of the department, a reduced cleaning effort was validated to optimize the turnaround time of the suite within the defined performance criteria. As a result, a less stringent cleaning methodology was developed and was found to effectively clean down to 1 $\mu g/m^3$ or less of residual contaminant during aggressive air monitoring. This, in return, produced a cleanup time of 20–25 minutes as opposed to 40 minutes for the earlier more stringent cleaning method. The benefit of the studies was an extension of the use of a midpotency dispensing suite capability to high-potency capability through enhanced cleaning procedures and without additional capital investment.

IV. DISPENSING OPERATIONS (ca. 1993)

A. Background and Scope

The purpose of this study was to develop contained pharmaceutical dispensing practices and hardware with accurate, repeatable, and documented performance data. Figure 5 shows a diagram of the setup. Over a period of six generations of containment design using flexible materials, data-based decisions were made on the efficacy of the containment systems.

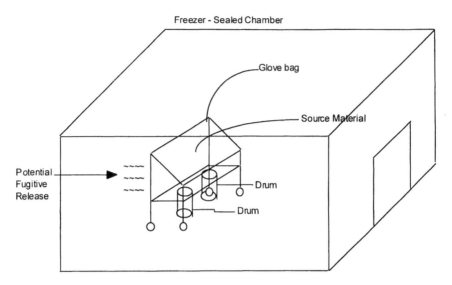

Figure 5   Dispensing operation setup.

B.   Method

Testing was done in a closed chamber—specifically a freezer room that was not
in use—with a hermetically sealed door and no active ventilation. Drums of a
surrogate material were connected to a dispensing glove bag and dispensed by
an operator. During dispensing, the concentrations of surrogate inside the glove
bag, the local concentration outside the glove bag, and the gross contamination
buildup* within the freezer† were determined using cassette samples at approximately 2 L/min. flow rate. Isokinetic sampling considerations were not adopted
during these studies. An indication of relative performance improvement as progressive design iterations were tried is shown in Figure 6. Since the inside-con-
tainment (the contaminant source) and outside-containment (operator breathing
zone) sampling methodologies were directly equivalent, it was presumed that an
appropriate containment factor§ of the containment device, in combination with
the operator's procedures, was obtained. Some hint that this was not the case

---

* The gross contamination buildup was monitored using randomized samples located in homogeneous
  (as distinct from local to the potential source) areas within the freezer.
† AHERA protocol for clearance monitoring referenced in Chapter 5.
§ Containment factor is a measure of the difference in concentration across the isolation barrier (i.e.,
  the glove bag). It is based on the same principle as a protection factor for a respirator, but it is not
  to be confused with the same.

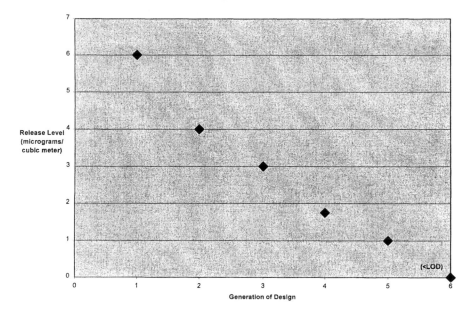

Figure 6 Performance improvement—release levels trending downward.

was indicated when a RAM 1 real-time OPC was used to monitor fugitive releases around the seams of a dispensing glove bag.

C. Results

While the statistically random samples around the glove bag were consistent, a pinpoint fugitive release was identified using a moveable fast feedback instrument. Transient increases in particle count, represented as mass changes by the RAM 1, were observed and assignable to a specific cause, which was a pinhole leak in the sealed edge of the glove bag. This only occurred in one glove bag out of a total test population of nine glove bags.

D. Conclusion

The major error in experimental design was due to insufficient dispensing cycles for each glove bag to achieve finite measurements on the homogeneous samples inside the freezer. While containment factors in excess of 1000 were observed based on LOD (Limit of Detection) interpretations, the containment factor might well have been greater than this had sampling been pursued longer and for more cycles. During later studies, cassette samples were placed at release points identified using the OPC and a more robust database was achieved that provided capability to identify minor leakage points. The overall benefit was a robust dispensing system having operator buy-in through their input into the design and procedures. The dispensing suite capability was extended to high-potency material handling without major capital investment as a result of performance-based measurements.

V. NEW FACILITY COMMISSIONING (ca. 1993)

A. Background and Scope

A facility was constructed to manufacture a high-potency drug material. The layout for the contained zones, including room pressurizations, is shown schematically in Figure 7. A filter dryer was contained within a dedicated isolation suite on the second floor and a packaging glove box isolator was contained in a dedicated isolation suite on the first floor. There was a small opening between the floors, which could cause potential cross-migration of any contaminant released. Each of the suites was connected to a decontamination shower having its own air supply and HEPA-filtered exhaust, and these connected to a gowning area. The gowning area connected with adjacent areas of the production facility. The gowning areas were established as positive pressure isolation zones. Both showers were at slightly negative pressure with respect to ambient pressure, and the processing areas were increasingly negative (-0.05 in wg). The study was designed to monitor the overall performance of the processing hardware, procedures, and facility design.

B. Method

The sampling system was set up based on the AHERA clearance monitoring concept (reference the AHERA guidelines) such that the contiguous processing areas represented one homogeneous zone. The gowning areas, having a common air supply, were also created as a single homogeneous zone. The production areas and gowning area were divided into equal area spaces for purpose of establishing statistically random samples, exclusive of the fugitive source areas in the processing room. Closed-face cassette samples using personal sampling pumps were taken at each of the random locations during each of the operations. Personal and pseudo-personal breathing-zone samples were taken when operators were present in the room or performing operations such as sampling and packaging. This array provided information on the initial background in all areas; potential and actual personnel exposure; any incremental backgrounds due to operations; and actual excursions during operations.

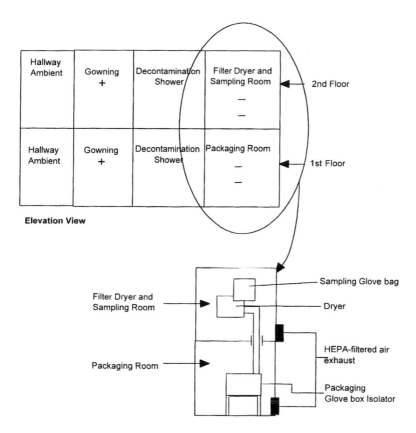

Figure 7 Oncolytic facility containment zone.

C. Results

During validation lots, no finite measurements were observed (the detection limit was 20 nanograms per filter, representing 0.01 μg/m^3 airborne concentration for the duration of a 12-hour sample). In addition, settling plates were located at each of the statistically random locations at the start of production. After 30 days, the settling plates were removed for analysis and replaced with clean settling plates. Two finite measurements were identified: one in the filter dryer room (0.16 μg) and one in the interconnected gowning area (0.04 μg). Follow-up cassette sampling at these locations returned finite measurements of 0.12 μg/m^3 and 0.03 μg/m^3, while all other samples were below LOQ. Observation of the process exposed the fact that the protective booties, required for quality purposes, were accumulating a very small level of contaminant through polishing the floor in the suite as the operators worked in the area. This contaminant was released from the surface of the booties as they were "snapped" off the operators' shoes and placed in the drum located in the gowning room by the entrance to the decontamination shower door. This release was detected by the samples located adjacent to the waste drum. This procedure was changed by placing the disposal drum in the decontamination shower area. No subsequent migration of drug material was observed for the duration of the 9-month campaign. Use of the fog generator to track ventilation airflow patterns in the filter dryer room showed that air entering from the ceiling-mounted air supply jetted to the floor in a band of less than 2 ft. in width. The air then proceeded to flow toward the HEPA filter intakes on the other side of the room by sweeping across the floor in a well-defined band of approximately 2.5 ft. in height, well below the level of the processing equipment (dryer and sampling glove bag). The remaining air in the room was stagnant (approximately 70% of room space).

D. Conclusion

The combined use of settling plates, air sampling, and ventilation-flow observation allowed us to focus on the assignable causes of the release and subsequent tracking by operators. Since sampling was conducted at a height of 5 ft. above ground, any small releases occurring during sampling (using the glove bag) were entrained into the stagnant zone, resulting in coating of surfaces in that zone, including the settling plate where the 0.16 μg deposition level was observed. Following room cleaning, retraining of operators, and securing a new glove bag around the sampling port, no measurable levels of active compound were found on the settling plates throughout the rest of the campaign. The overall sampling campaign allowed building supervision and operators to identify extremely low-level sources of contamination and tracking.

VI. PERFORMANCE QUALIFICATION OF A SEEDING ISOLATOR (ca. 1994)

A. Background and Scope

A glove box isolator was used to achieve collection and delivery of seed material using a dedicated seed isolation chamber. The seed isolation chamber was mounted inside the glove box and filling and dispensing were achieved using a two-piece mated docking port located on the top of the glove box. The seed container was manipulated only from inside the glove box, with no intentional leakage from the seed container into the glove box. The scope was to show whether there was full closure between the seed container and the internal glove

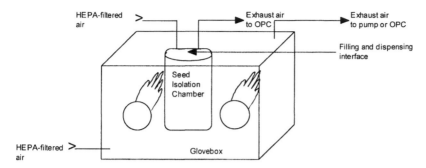

Figure 8 Seed container mounted inside glove box.

box space. Figure 8 shows the schematic test setup to measure the extent of isolation between the seed container and the internal glove box space. A particle counter was set up to continuously monitor the smoke level inside the seed isolation chamber. A sample pump was set up to pull air through the glove box at the same flow rate as the seed isolation chamber. A particle counter was used intermittently to measure the decay rate of smoke levels inside the glove box.

B. Method

The instrument used was an OPC (RAM 1). The clean air in the glove box was contaminated by infusing smoke from a theatrical smoke generator, bypassing the HEPA filter, until the chamber was filled with smoke. Monitoring of the seed isolation chamber was initiated as soon as this procedure was complete. The pump for the glove box was also turned on to draw air out of the contaminated chamber and blow it into the air exhaust for the room. In this way fresh air was being drawn into the contaminated chamber at a known rate through a HEPA filter. The operator monitored the seed container continuously, and intermittently monitored the concentration inside the glove box–contaminated chamber until such time as the instrument reading decreased below the overload level to produce finite readings, at which time the reducing concentration in the glove box was monitored for a period of 4 hours.

C. Results

Results of the study showed the seed isolation chamber had no detectable migration of material across the boundary from the contaminated zone to the clean zone of the seed chamber. The glove box internal volume was approximately 8 cu. ft. and the seed chamber volume was 1 cu. ft. The seed chamber was much more sensitive to change than a glove box, while the glove box provided a stable insult challenge of smoke.

D. Conclusion

Not only was no migration of contamination across the seed-isolator boundary observed, but the decay of aerosol in the contaminated zone of the glove box followed a classical exponential decay. The containment factor across the boundary was shown experimentally to be in excess of 5 million based upon extrapolation of the glove box decay trend back to the start of the test (time zero). The study was repeated by three operators to complete the performance qualification (PQ) for the hardware. An additional advantage gained was buy-in from the operators by providing a test protocol having real-time data such that they could observe the true protection provided by the system. By introducing the smoke into the glove box without contaminating the seed isolation chamber, quality control concerns were also mitigated. The benefit was proven performance of pharmaceutical process isolation from a quality control viewpoint, and proven shirtsleeve work environment for the operators.

VII. RIGID ISOLATOR (GLOVE BOX) PERFORMANCE TESTING (ca. 1995)

A. Background and Scope

The scope was to conduct performance testing of rigid isolator using the full AHERA-based statistical sampling strategy. Rigid isolators (glove boxes, in this case) were subjected to a number of different usage cycles typical of those to be introduced to a facility after construction was completed. Specifically, the study was to measure the performance of the rigid isolators during charging, packaging, and transference activities.

B. Method

In this case a combination of personal sampling pumps, high-flow pumps, and a sampling manifold (reference Chapter 5) were used to collect a total of 23 samples per test run.

Sampling setup was again based on the AHERA principle. The isolator was measured, divided into equal area volumes, and extended outwards by one volume, in each direction, on the x-, y-, and z-axis as shown in Figure 9. That is, the number of sample zones went from 12 of equal volume on the inside to an additional 68 zones surrounding the glove box. The points of fugitive release—at

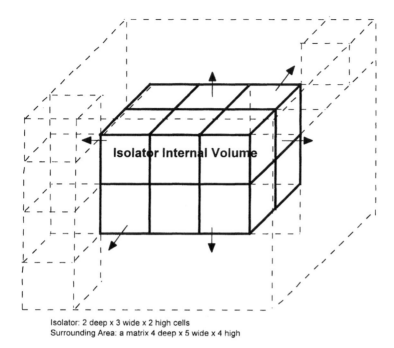

Isolator: 2 deep x 3 wide x 2 high cells
Surrounding Area: a matrix 4 deep x 5 wide x 4 high

Figure 9 Dividing isolator into sample cells.

make and break points above, below, at the box's sides, and at the gloves in front of the glove box—were then excluded from the sample point matrix. The remaining external zones were numbered sequentially. Using a random number generator, five discrete sample locations, or zones, were selected from all those around the box. At a radius of 15 ft. from the center of the box (the homogeneous surrounding area in a potentially contaminated zone 30 ft. in diameter), 24 equal area locations were created and sequentially numbered. Again, a random number generator was used to identify five locations. Two entrance doors were located in this facility, and since both provided supply air to the room, they were both assumed to be homogeneous sources of relatively uncontaminated air. These locations were used to create 24 additional sampling zones for supply air. Again, a random number generator was used to identify five locations. Cassette samples were taken inside the isolator (at the source) at the east and west sides to provide a measure of source variability, the five random locations adjacent to the box, the five random locations at the 15 ft. radius, the five random locations simulating supply air, two pseudo-personal samples at the operators' breathing zone at the glove box, and conventional personal samples on the operators. Figure 10 depicts the sampling layout.

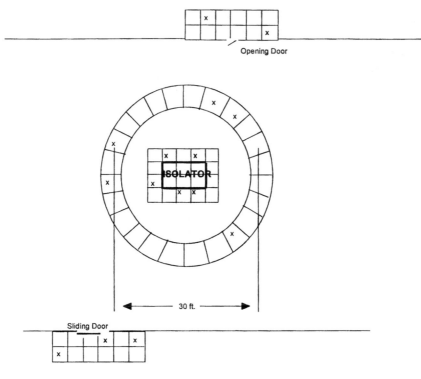

Figure 10   Overhead view of overall statistical matrix for assignment of sample locations.

C.   Results and Conclusion

From a containment perspective, this provides a measure of 1) containment factor across the isolator using the two internal source samples and the five isolator surround samples; 2) fugitive release to the general environment using the 15 ft. radius sample locations; and 3) the contribution to the background due to supply air. Additional high-volume samples, using an open-faced cassette and a high-volume pump placed at disconnection points, were used to measure any possible fugitive release due to assignable point sources. These cassettes were changed with each iteration of the process, such that discrete measurements were obtained for each cycle and operator. Additional personal data was obtained using personal sampling pumps and cassettes on the operators during the operations. Pseudo-personal samples were generated by placing closed-face cassettes against the front of the isolator in the hypothetical breathing zone of the operator and sampling, using the manifold, for the full duration of the operations. In this manner, the operator potential exposure due to the hardware and procedures was measured without the contribution of operator contamination of clothing. Using this strategy, personnel contamination was compensated for by virtue of the pseudo-personal samples. The inside of the isolator showed progressive build up of contaminants with increased number of operations. The outside of the isolator showed no isolation release either randomly or in the breathing zone. Fifteen-feet perimeter samples and supply air samples were immeasurably low. Personal and make/break connections showed finite measurement, and characterized the weak point in the system. In total, the studies led to a better understanding of the weak spots in hardware design and operation, with appropriate remediation. The longer term gain was the confidence to start up a new containment facility as a shirtsleeve work environment.

VIII.   OUTDOOR PROCESSING (ca. 1995)

A.   Background and Scope

A purchased raw material was newly classified as a high-potency material. No available sampling methods could be found nor developed in the time available. A containment solution for charging the material into an outdoor reactor was developed but performance data on the solution was desired in order to make both a facility capability and an appropriate personal protective-equipment evaluation.

B.   Method

The only tool available was an OPC having sensitivity down to the ng/m^3 range—preferably with isokinetic sampling. The only instrument available for this purpose was the CLiMET series of OPCs. An algorithm was developed to convert particle counts to approximate mass concentration based on the median mass and density of the particle size ranges. With only two instruments physically available for the study, it was necessary for the investigator to keep relocating the instruments in the selected random sampling array (reference AHERA guidelines) in order to obtain the data.

C.   Results and Conclusion

Over a period of nine lots of product, the operators showed an initial capability to operate below the allowable exposure guidelines. A failure to meet the guideline occurred as they attempted to perform the operation more quickly. With the real-time observations provided by the OPC instrument, the operators were persuaded to improve their procedures and restore the operation to a fully controlled environment with no fugitive releases during the charging operation. The ability to change the work practices with the immediate data feedback that real-time technology provides created a new tool for application in an Occupational Health and Quality Control environment. Additionally, the environment was protected by adoption of improved procedures, since fugitive releases from the process were not a permissible occurrence. Most important, however, was the operator buy-in to both the procedural changes needed and the value of the real-time data used to provide performance feedback. The study added new insight into how far flexible-barrier technology can be pushed to adapt existing facilities for handling high-potency materials.

IX. OPERATIONAL QUALIFICATION OF A SAMPLING GLOVE BOX (ca. 1996)

A. Background and Scope

Sampling glove boxes of a new design were delivered to a production site. Before beginning Performance Qualification (PQ), Operational Qualification (OQ) was performed using smoke as a challenge. Rapid empirical measurement of smoke leakage, made possible by real-time air-monitoring techniques, was used to identify problems before extensive quantitative testing was conducted.

B. Method

A glove bag was built around the sampling glove box. The bag was filled with smoke from a theatrical fogger. Visible leakage occurred into the sampling glove box through the door gaskets. After replacing the glove box gaskets, the test was repeated. Subsequent repairs were carried out until no visible leakage occurred into the box from the surrounding glove bag interior. At this time PQ was conducted using the same strategy as in the Seeding Isolator Study with the enclosing glove bag providing the smoke challenge and the inside of the sampling glove box as the sensing chamber.

C. Results and Conclusions

The gaskets that were received from the manufacturer were too rigid, resulting in leakage of fog into the glove box. After replacing the gaskets with a softer material and adjusting the tensioning on the door locks, both the visible and quantitative testing showed no leakage of fog into the glove box.

Acceptance of the glove box as delivered would have resulted in serious containment issues during production. As a result of empirical and quantitative testing, validated using real-time monitoring methods, glove box performance was improved, and the units were able to be put into service in a shortened amount of time. The efficacy of the test procedure proved to be a significant step toward preventive maintenance. As in the previous case, once the sample glove box is cleaned, and shown to have no measurable leakage during testing, the glove box is immediately ready for use without concern for loss of containment. The study mitigated contamination of an entire processing suite at the OQ stage of testing where hardware failure was documented and then remediated. As a result, expedient solutions were adopted and further tested to ensure that processing was contained and achieving a rapid closure to the OQ process.

X. FULL FACILITY COMMISSIONING DURING EXTENDED PRODUCTION (ca. 1997)

A. Background and Scope

A newly constructed four-story building containing 17 dedicated isolation suites for manufacturing high-potency drug material was intended to be operated in a fully shirtsleeve mode. A sampling plan was required to assist the building owners to control the facility, maintaining the shirtsleeve environment throughout normal operations.

B. Method

The AHERA strategy was used to establish the sampling plan. Each production area had its own group of statistically random samples exclusive of fugitive source locations. Four stories of public areas were considered homogeneous areas because they were all provided with outside air and were represented by five sample locations. The sampling plan included active air sampling using cassettes for sample collection, in combination with battery-powered personal sampling pumps and a high-flow manifold system. Settling plates were used for periods of 7, 14, and 28 days of monitoring, depending on the circumstances in the building. On one occasion multipoint sampling was performed with an 18-channel real-time optical particle counter in a packaging room.

C. Results and Conclusions

We were able to maintain continuous operational control in the packaging room by observing no quantitative measure of contaminant for a period of 7 months until an incident occurred in the packaging room. In that incident, a loss in weight hopper separately generated a fugitive release incident, which was noted by an instant change in the OPC readout, remaining elevated for a period of 18 minutes. Chemical assay of the associated cassette sample returned a value of 3 ng/m^3 for a 12-hour TWA (time-weighted average) event, almost 100x below the effective exposure level guideline. This was the first time that full-time OPC monitoring had been used to monitor a process. Although there was no visible release, we were able to immediately evacuate the room with no measurable exposure to the operators. This would not have been possible using conventional sampling methods. The continuing studies enabled building personnel to control the fate of the facility.

XI. PARTICLE CONTROL TEST COMPARISON (ca. 1997)

A. Phase I

1. Background and Scope

The scope of this study was to establish whether the American Society of Heating, Refrigerating, and Air-Conditioning Engineers (ASHRAE) 110 test procedure using sulfahexafloride (SF_6) vapor truly represented equivalent behavior of particles during performance testing of lab hood enclosures, glove boxes, glove bags, and reverse-flow isolators. A particle effusion system, having flow characteristics the same as a SF_6 diffuser but using a particle-generating mill as a source, was developed and used in the ASHRAE test. The SF_6 diffuser and the particle effuser sources were used independently to test the performance of the different hardware under identical-use conditions.

2. Method

The hoods were initially tested and performance rated using SF_6 and were adjusted until they met all the ASHRAE criteria. The particle effuser was located in the hoods at the same location as the SF_6 diffuser and lactose dust of <3 μm particle size was emitted into the hood at a delivery rate of 4 mg/m^3. Lactose was used because of the extreme sensitivity of the lactose assay (0.01 μg per filter LOQ and 0.002 μg per filter LOD). Figure 11 shows the sampling set up. Sampling was performed at breathing zone height at the hood face. Samples were taken on the mannequin adjacent to the intake position for the SF_6 probe and the three additional positions shown in the figure. As shown in Figure 11, a parallel set of samples was placed 6 ft. distant from the hood entrance in order to evaluate artifact data due to background concentration from room contamination and the particle generator unit (located inside a glove bag in the room). Side-by-side sampling, using real-time OPCs, was performed at the following locations: three at the left, center, and right hood positions and one placed 6 ft. behind the mannequin. Three replicate tests were conducted for a duration of half an hour for each test once stable dust flow was achieved. A laser light beam was used to observe the dust effusion. The cassettes were run at a flow rate of between 7 and 10 L/min. using an 18 mm–cassette face opening to approach a pseudo-isokinetic mode of sampling and to improve the analytical sensitivity.

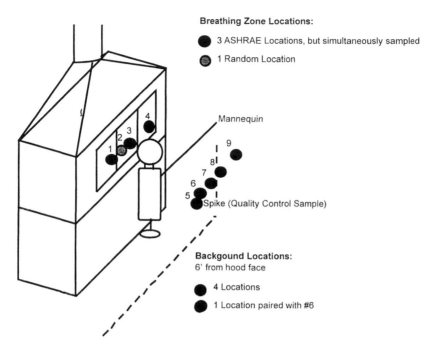

Breathing Zone Locations:

● 3 ASHRAE Locations, but simultaneously sampled

● 1 Random Location

Mannequin

Spike (Quality Control Sample)

Backgound Locations:
6' from hood face

● 4 Locations

● 1 Location paired with #6

Figure 11 ASHRAE 110 study—mannequin setup.

3. Results

No detectable levels of lactose were found for any of the cassette samples, either at the hood face or in the room. Transient excursions were seen at the mannequin breathing zone by the particle counters whenever the hood was approached to observe the performance of the effuser and whenever a hand and arm entered the hood space to adjust the source cassette located behind the dust effuser. Particle count excursions were directly attributed to the approach of operators toward the hood and when backing away from the hood. In other words, any operation effecting possible turbulent flow at the hood face caused spontaneous particle release out of the front of the hood, but was below the LOD of 0.002 μg/m^3 using a cassette filter monitoring system.

4. Conclusion

Because of the lack of quantitative measurements from the cassettes, it was not possible to calibrate the OPCs for equivalency of mass conversions. Nonetheless, the observations become significant when the effect of operators and equipment is introduced at the hood face (i.e., immovable and moving obstructions).

B. Phase II and Phase III

1. Background and Scope

Concern was expressed about the possibility of coating of surfaces at the front lip of the hood and on the working areas due to dust deposition, static charge, etc. This test was designed to address those concerns by measuring deposition at these locations.

2. Method

In order to measure the coating effect, 60 mm–diameter glass settling plates were placed beside the effuser on 6 in. or 9 in. centers across the hood face, depending on the width of the hood opening, as shown in Figure 12. An additional set of settling plates was placed midway between the hood face and the plates just described. The settling plates were left in place during the full series of three half-hour tests and then removed for chemical analysis.

3. Results

Chemical analysis of the settling plates is less efficient than cassettes by a factor of 10, but finite measurements were accomplished depending on the hood and

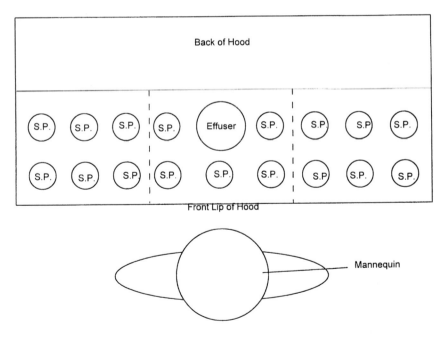

Figure 12   Settling plate sample setup.

the test conditions. A visible surface haze surrounded the settling-plate location after the settling plate was removed.

4. Conclusion

While SF_6 (ASHRAE 110) testing showed equivalent breathing-zone measurements, it is not able to show the long-term effect of dust accumulation due to minute but repeated releases and their effect on cross-contamination. These tests indicated that placebo particle testing is a more realistic way of predicting the performance of fume hoods relative to personal contamination through contact with contaminated surfaces than the ASHRAE test. This is due to the challenge agent being a solid rather than a volatile agent having no potential for residual surface contamination. The observations from this study challenge many of the accepted views of safe hood usage based on vapor testing alone. Vapors are continuously drawn away from the surfaces whereas particles settle on the surfaces creating a source of contamination.

XII. PPE PERFORMANCE TESTING (ca. 1997)

A. Background and Scope

It was desired to use PPE with an extremely high protection factor. Due to the extreme dynamic range anticipated between the containment source and inside the protective suit, it was not possible to use the same monitoring technique without significant design modification.

B. Method

A submicrometer aerosol fog was generated using a commercial fog generator. Because of the extreme sensitivity of the real-time OPC, which was capable of seeing nanogram concentrations of aerosol fog, it was necessary to dilute the challenge concentration of fog by ratioed dilution with HEPA-filtered air over 10,000 times before presentation to the OPC for analysis. Dilution ratios of samples were adjusted such that they were all presented to the analyzer within the linear calibration range of the sensor. Figure 13 illustrates the general concept.

C. Results and Conclusion

The ultimate sensitivity of the OPC after serial dilution was 200,000-fold below the challenge concentration of the aerosol source. Repeated testing of the PPE (ILC Dover Chemturion suit having a gas tight closure over a zippered closure) was conducted. The concentration inside the suit was below the lower detection limit at all times. A protection factor of greater than 200,000 was obtained for

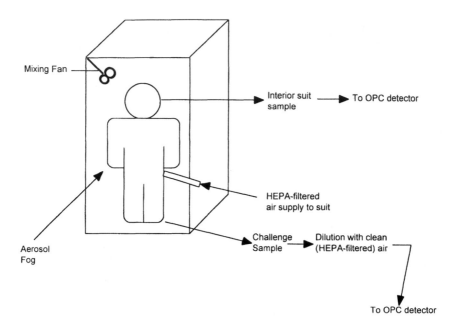

Figure 13 PPE performance testing schematic.

the suit. This is a significantly greater protection factor than was needed, but the suit's comfort and ease of use encouraged operator buy-in to its effective usage. An added bonus was the significant financial gain in providing an overprotective suit that turned out to be a reusable item based on cleaning validation.

XIII. CONTAINED PACKAGING SYSTEM—INITIAL DEVELOPMENT (ca. 1997)

A. Background and Scope

A collaborative study was undertaken to develop a contained packaging system for a rotary vacuum dryer (RVD).

B. Method

A mockup of an RVD was fabricated and the system was charged with 450 kg of lactose. A team of engineers and pharmaceutical operators packaged out of the mocked-up RVD using a prototype system. The performance of the system was monitored using personal monitoring cassettes and fixed-location cassettes inside a 6-by-6 foot plastic enclosure located below the mocked-up RVD. Additional monitoring was performed using an OPC to indicate transient incidents. The overall goal was to achieve contained packaging while not exceeding an exposure potential of 1 $\mu g/m^3$, the time-weighted average for a 12-hour work period.

C. Results and Conclusions

During the earliest trials, it was demonstrated using the particle counter that conventional ultra low density polyethylene (ULDP) film was not an appropriate material of construction for the packaging container due to its fragility when subjected to any kind of stress (e.g., bending, flexing, heat sealing). A flexible polyurethane film was demonstrated to be far superior and created a benchmark for the desirable physical properties of a barrier. The conceptual and prototype packaging system used in conjunction with a polyurethane packaging container met the 1 $\mu g/m^3$ criterion. These results, and subsequent procedural changes, were greatly expedited by data provided from the OPC. This in turn has impacted other projects as well as this collaborative project. The system has since been incorporated into normal operating practice.

7
Selecting the Correct Technology

Hank Rahe
Contain-Tech, Inc., Indianapolis, Indiana

I. INTRODUCTION

This chapter provides a systems approach, including tools and methods, for understanding and selecting the proper solution to a variety of containment problems that occur in all phases of the life sciences. The technologies described represent engineered solutions, that is, solutions not dependent upon personal protective equipment to achieve a defined level of protection for individuals working in the area in which potent compounds are present. There are a number of different technologies available to provide the proper level of protection for individuals handling potent compounds, and these technologies—combined with other tools described in the chapter—help to navigate the path that will result in a successful project.

The hierarchy of containment technologies shown in Table 1 is a guide to the selection of which technologies should be considered as a potential solution. The criteria for selection of the technology should be based on the quality of airborne material that is acceptable without exceeding a predetermined amount, typically based on an exposure limit. The exposure limit defines the level of exposure that is acceptable without impacting the health and safety of individuals working in the environment. The hierarchy of technologies defines which of the technologies can deliver a feasible solution. In some cases, there are several technologies that may provide acceptable levels of protection, allowing an economic analysis to determine the most cost-effective approach to a solution. The elimination of the technologies, which cannot deliver adequate protection levels, saves time in investigation and focuses efforts where there is a high probability of

Table 1 Hierarchy of Containment Technologies

Technology	Level of control
Barrier	Less than .1 $\mu g/m^3$
Directionalized laminal flow meter	Approximately 10 $\mu g/m^3$
Laminar flow meter	Approximately 30 $\mu m/m^3$
Local or point exhaust	50–100 $\mu g/m^3$

Note: The level of control described for these technologies is approximate. Individual powder characteristics and systems design will impact the ability to contain potent compounds. Individual systems should be tested and validated to determine their ability to control airborne potent compounds.

success. For example, it is probably not worthwhile to consider a local exhaust system approach when the level of containment required is 1 μg of the potent compound per cubic meter of air.

Containment problem solving involves more than selecting the correct technology. It requires a process that includes the following elements: definition of the problem or project, consensus upon the definition, selection of the appropriate technology, and integration of the technology with the process. Understanding the physical tools, technologies, and capabilities is not the most critical factor to project success. The most critical factor is understanding the goals and communication of these goals during the life of the project. To achieve this, a tool which I call the *requirements document* is used both for internal and external communication. The internal communication is for both the current team members and any new member that will join the project in progress. For current members it is a diary that describes the goals and objectives of the project, and for new team members it is the history and written record of decisions taken and the defined goals and objectives set by the team. Many project teams are sidetracked by retracing the steps for the new member whose contribution should be focused in moving forward rather than revisiting the past decisions.

The requirement document clearly describes the activity or process, including all physical interactions, that will result in possible exposure. The document will include descriptions of process activities, sampling, and routine maintenance. Where multiple shifts are involved, the requirements document should be reviewed and agreed upon by each of the shifts to assure total agreement. This description will detail the process that is occurring within the containment zone, clearly defining where an interaction is needed that will require a person to perform a task that could result in potential exposure if proper containment is not in place. Areas of particular interest are taking of samples and places where the containment will be broken when the process is completed. It is important for the team to create a written document that describes the agreed-upon process or activity before involving vendors of containment devices. A properly selected project team has the most complete understanding of the entire process and can use the document to describe the physical interactions that must take place between the process and the individuals operating the process.

The major challenge faced with any successful containment project is to integrate the containment with the process. The major reason for failure of containment projects is the lack of integration of process with the containment solution. The technology may be completely acceptable but if the integration has not been completely thought out in terms of ergonomics and ease of operator interface, the project probably will fail. Do not assume companies that happen to provide containment equipment understand your process, even though they have worked on similar projects.

Containment of a potent compound should reflect a systems approach, with the system having at least two of the following four levels of protection in place. The four levels of protection for a containment system are the internal environment of the containment device, the device itself, the external environment in which the containment device is placed, and the structure housing the containment system. An example using barrier/isolation technology would be the pressurization scheme inside the barrier/isolator, the physical structure of the barrier/isolator, the pressurization scheme in the room containing the barrier/isolator, and the wall of the room.

Understanding the basic components or elements that make up the various containment technologies is important. The knowledge can be gained from several sources including consultants and containment-equipment vendors. The reason this information is important is that each component or element offers a variety of choices that can impact can both the cost and function of the system. Using the technologies described in Table 1 we can review the components that make up the technologies and the alternate choices that impact cost and function. It is important to remember our baseline for choices is only those technologies that result in acceptable containment levels based on allowable airborne concentrations of the potent compound.

II. BARRIER/ISOLATION TECHNOLOGY

Barrier/isolator systems are made up of four basic parts: the physical structure; the internal environment; the transfer and interaction technologies; and the monitoring systems. Within each of these parts there are a series of choices that will greatly impact the success of the containment project.

A. Physical Structure

Physical Structure of a barrier/isolator is generally classified by the appearance of the structure and falls into two categories: hard shell and soft shell. The next level of differentiation is the materials of construction, with the choices for the hard shell being plastic, plexiglass, glass, and/or stainless steel, and the choices for the soft shell being a soft plastic or otherwise flexible film. Important issues concerning the construction of the shell focus on three issues, durability or integrity of the system, ability (or need) to have cleanable surfaces, and visibility of internal operations. Selection of which alternative will best fit the application should be reviewed with these criteria in mind. The durability or integrity of the barrier/isolator shell should be consistent with the environment in which the unit will be used; the process and tools involved in the activity; and the frequency of use.

The cleanability issue involves the quality of finish, welds, joints, and corners. The finish quality is a balance between a surface smooth enough to allow the removal of materials (using, for example, a clean-in-place type of spray-ball action), and the issue of reflectivity of light, which can cause operator eye stress. A reasonable compromise can be a number four polish. The use of an electro polish surface will create too much reflection for viewing operations.

The criteria for welds and joints differ with the past traditional use of barrier/isolation technology, and requires attention in specifications when ordering a unit. Manufacturers of units understand the need for quality welds and testing for complete penetration, but not always the concern for cleanability of these pieces of equipment. For common understanding, obtain sample welds that are "ground smooth and free of pits" from the vendor and agree to acceptable quality standards before fabrication. Joints around penetrations such as viewing areas, air inlets, and exhausts, and other required components should be smooth and free of ledges. The mounting components used for attaching the glass or plastic that make up the viewing areas have improved. Yet, close attention to detail is required to assure that surfaces are smooth. Materials used for gasketing need to be checked for durability as well as product compatibility. Effects of exposure to cleaning agents should also be checked.

Visibility is the other important consideration. Pharmaceutical operations require a number of interactive steps, including visible checks of activities. Physical mock-ups of the barrier/isolator should be constructed of solid materials to determine interactions both physically and visually. If possible, the actual machine or operations should be used in this type of evaluation, with both operations and maintenance personnel involved.

The physical shell defines the boundary of the barrier/isolator and determines the points where interaction is possible. The complete understanding of interaction is a different concept for pharmaceutical operations and, in many cases, a major problem in implementation. Front-end planning and definition of interaction requirements and constraints are required for the successful project. Penetrations of any type represent a potential weak point in the system and should be incorporated only if needed, rather than on a nice-to-have basis.

B. Internal Environment

The major paradigm shift required from past traditional design is the amount of airflow needed within the environment. Traditional systems are based primarily on negative pressure rooms, with air changes from fifteen times per hour up to sixty or greater. These parameters have been carried over to the design requirements for barrier/isolators oftentimes by inexperienced individuals. The negative pressure of a barrier/isolator is a form of secondary containment to the system and should be designed to allow for protection from a breach of the barrier. The most common breach is a glove port or bag ring not properly sealed or torn during operation. A normal design criterion, developed by the nuclear industry, defaults to the next level in the hierarchy of containment technologies: laminar flow. A typical design requirement incorporates a flow velocity of 120 ft./min. across the breach. It is important to remember that the physical structure is designed and tested to have integrity without the negative pressure. Both initial testing and continued certification of the units should be based on tests with positive pressure.

The internal airflow of a barrier/isolator designed for containment should allow clear viewing of the operation and removal of any heat generated by the process equipment contained in the system. The decision to recirculate or throw away the air is a function of any special atmosphere requirements. If low humidity or inert atmosphere is required, the system should be recirculated to reduce cost. The exhaust requires filtration to capture the potent materials before release of the airstream.

The airflow system's size is determined many times by the flow requirement of the emergency or breach volume requirements. A number of systems have been developed using two-stage fans, having speeds for normal operation and emergency situations. Emergency-exhaust design for recirculating systems require ducting to an outside vent after filtration. It should not be assumed that barrier/isolator manufacturers understand the previously mentioned information. The pharmaceutical industry differs from the nuclear or electronics industries many vendors are more familiar with; in pharmaceutics, much larger quantities of powders are handled.

Lighting and utilities also require planning. Lighting should be located exterior to the physical structure of the barrier/isolator with see-through panels allowing the light to enter. Utilities can be provided interior to the barrier/isolator through several means, including fixed panels inside or flexible connection ports. Attention should be paid to potential electrical hazards that could be present from solvents or dust.

III.   TRANSFER AND INTERACTION TECHNOLOGIES

A.   Transfer Technologies

Transfer of materials between a barrier/isolator and the outside, without exposure of the potent material, is one of the biggest challenges of any system. This section identifies the available technologies and discusses the applications of each for handling potent compounds (Table 2).

For this discussion, the term *soft shell* refers to flexible containment devices that are generally used over longer-term applications, for example a sterility test unit. Glove bags are typically used and disposed of shortly thereafter. Physical differences between the two are the thickness of material (greater in soft-shell applications), and sophistication of transfer provisions (higher in soft-shell units).

1.   Double Posting Ports

This technology is the mainstay of high-integrity transfer technology. The double-posting port design and several variations are manufactured by a number of vendors. The idea of this component is a double door, capable of separating to form two seats for separate enclosures, allowing both to maintain integrity. Developed for the nuclear industry, the design has proven transferable to the pharmaceutical industry. The advantages of the double-door posting port for transfer of materials, is a high level of containment integrity and a positive sealing of two enclosures for the transport of materials between operations. Disadvantages are the durability of the rotating seals, the rotation required for docking of most systems, the round configuration of the door by most designs, and a small amount of contamination sometimes left on the seal ring after use.

Table 2   Transfer Technology Available for Barrier/Isolators

Transfer technology	Hard shell	Soft shell	Glove bag
Double posting port (RTP)	yes	yes	no
Bag rings	yes	yes	yes
Air lock	yes	yes	no
Airflow	yes	no	no

2. Bag Rings

This transfer technique involves the use of plastic sleeves or bags to introduce materials into or out of a barrier/isolator. The method is more procedure dependent in sealing technique but has proven effective in handling potent compounds. Bag rings can come in a variety of sizes and shapes to meet requirements and offer flexible means of transfer. Two disadvantages are longer-term dependability of seals and working with materials that can be torn or cut by sharp objects. A major advantage is that the sealed bag or sleeve can act as a low-cost transport device for exposed tools for change parts.

3. Air Locks

Air locks are defined as areas that act as transfer or transition spaces between independent areas. The area inside the air lock becomes a buffer for each of the adjacent areas, so each can have materials transported inside without exposing the other area. If potent compound exposure or even some processing is taking place in the air lock, the exposure of the area would require a deactivation cycle take place before opening the air lock to the outside area. This approach has had application in aseptic operations for a long time and does offer the advantage of more area for solid transfer than the bag-ring approach. A disadvantage is that it does not provide a means of transport for the materials in a controlled environment, as does the double-posting port and bag-ring systems. This approach should be viewed as a one-way means of introduction into the barrier/isolator.

4. Airflow

The use of an air curtain as an integral external component of the barrier/isolator structure offers a means of introducing materials into the unit. This technique requires an extensive study of the airflow patterns and dynamics, especially while the air "wall" is being penetrated by a given object, to assure that potent materials do not escape the system. The use of this type of system for primary containment requires significant back-up systems for the air supply. It offers possibilities when utilized as a secondary containment approach.

B. Interaction Technologies

Allowing people to interact with the process or equipment contained in the barrier/isolator is an important part of pharmaceutical applications. Typical pharmaceutical processes require more interactions than industries using this technology previously (Table 3).

Table 3 Interaction Technology Available for Barrier/
Isolators

Interaction technology	Hard shell	Soft shell	Glove bag
Robotics	yes	no	no
Flexible membranes	yes	yes	yes
Half suits	yes	yes	no
Glove ports	yes	yes	yes

1. Robotics

This technology is discussed a great deal but applications in the pharmaceutical industry are limited. The limiting factors to the application include a lack of high volume, a high cost, and nonrepetitive movements. It is true that robots can be programmed for a large variety of activities, but this flexibility is expensive. Much of the process equipment includes automatic sampling systems but for most applications a personal interface is still required.

2. Flexible Membranes

This application provides flexible material that is designed to be pliable enough to "reach through" by way of stretching the membrane with no gloves. Although I have read articles concerning this approach, I have not experienced an application. Some concern has been expressed about the cleanability of the membrane and limitation of interaction with the inside of the environment.

3. Half Suits

The use of half suits was developed to increase lifting capabilities and expand areas of reach within the barrier/isolator. Initial designs involved coated fabric and a full-view, attached helmet. In one case, a disposable half suit, developed by a major pharmaceutical firm in conjunction with a manufacturer of personal protective gear, offered the advantage of being disposable, eliminating cleaning requirements. Disadvantages of the conventional half suit are the practicality of cleaning; difficulty in entering and exiting the suit; and personal hygiene issues by multiple users.

4. Glove Parts

This is the most commonly used technique for interaction. The sleeve-and-glove arrangement come in either one or two pieces. The one-piece offers slightly more containment integrity while the two-piece offers better fit for gloves with multiple users and less expense. Reach and weight of lifts are the important issues in design and selection. Most glove ports are located about fifty inches from the standing platform of the operator. The average reach is approximately twenty-two inches. These factors must be considered in any activity.

C. Monitoring Strategies

Monitoring of operations is critical for providing a safe workplace. Developing a plan or strategy for measurement of the levels of potent compound outside the containment systems is necessary. The measurement program should involve both facility and personnel monitoring. The current state of the market (i.e., what person can go out and buy) of measurement is such that online (or, real-time) analytical tools for monitoring chemical-specific solids are not practically available for most compounds. This lack of online exposure monitoring requires extensive, initial validation of containment systems and a continued operator-exposure monitoring program. This section outlines several approaches for establishing such programs involving both facility and personnel.

1. Facility Monitoring

Currently, state-of-the-market measurement tools for the determination of airborne particles are particle counters that can determine both a total count and the number of particles by size. Particle counters typically identify particle sizes in ranges from submicron (0.2 micron) up through 10 μm and higher.

This type of measurement determines total particle counts but requires a level of interpretation to determine the amount of actual potent compound present. This interpretation can involve identifying the percentage content of potent compound in the material and assuming that the materials are homogeneous. The design criteria for containment systems might be set by controlling the total particulate mass to the exposure limit guideline, thus allowing online feedback with (in many cases) a considerable safety factor.

There are a number of quality online particle-counter manufacturers offering state-of-the-art system components. There can be significant differences in level of service and support from each, depending on a project-by-project basis.

Particle counters use several different methods of counting particles and also use different collection methods. Of these two factors, the most important is the collection methods. There are two approaches to collection. First, a remote-reading device that collects samples from a number of locations by pulling a fixed volume of air through a tube arrangement to the reading device. It then counts the particles and sends the counts to a recording device. This approach can service a number of remote locations with a single reading device but has the disadvantage, if not carefully compensated for in the design, of allowing materials to collect in the tubing, which can lead to false readings. As a general rule, this method has typically not been recommended in the past for highly potent compounds.

The second approach works on the principle of reading at the point of sampling and sending a signal back to a central processor, which inteprets the information. This provides more accurate information but has the disadvantage of a number of remote devices called *bricks*, that require service and calibration. This is the preferred method for potent compounds.

Particle counters, as monitoring devices, should be located in areas where small differential counts are meaningful. An example location would be, the air duct downstream of the HEPA filters, where it can monitor the integrity of the filter system. In general, placing particle counters in areas having high particulate counts, such as the interior of the barrier/isolator or uncontrolled surrounding areas, offers little value if not a contributing piece of a larger, well thought out monitoring strategy. The background counts are too high to allow detection of the potent compound at the levels required.

2. Personnel Monitoring

Monitoring of personnel is a key element of containment. A qualified Industiral Hygienist is a required member of any project team developing a potent compound facility. The responsibility for the initial and ongoing monitoring of personnel falls to this individual's division of the company.

Current MSDS (Material Safety Data Sheets) must be available to all personnel having potential product. This document should be available from the raw-material supplier. A new, standard, sixteen part MSDS (ANSI Z440.1) was implemented in 1994 and should be used as a guide in generating new MSDSs. Reference for distribution requirements of MSDS for pharmaceuticals can be found in OSHA 29CFR Parts 1910, 1915, and 1926 (Federal Register/April 5, 1994).

3. Barrier/Isolator Monitoring

Following is a brief description of monitoring devices and their typical functions.

Pressure detection   Used to determine pressure difference. Units may be as simple as gauges with visual readouts or units connected to alarms. Some are connected into facility information and recording systems.

Air filter leak detection   Online feedback is important when dealing with very potent compounds. The pressure drop approach to monitoring filters is not sensitive enough to detect a leak of quantities that are above exposure levels. As mentioned earlier, online particle counters are recommended after the filtration system, to determine integrity of the filters for compounds having an exposure limit below 10 μm/m3 of airborne concentration. These systems can be integrated with facility information systems for record keeping and alarming.

Gas leak detectors   Instruments such as oxygen analyzers, can be used to determine proper processing environments and to act as a personnel alarming device. Gases such as helium can be used internally to the barrier/isolator to achieve detection in cases of extremely potent compounds. Integration with information systems for record keeping and safety should be considered.

D.   Directionalized Laminar Flow Technology

This technology is a combination of laminar flow and local or point exhaust, and was developed to improve the capability of typical laminar flow technology used in downflow booths. Locating a point exhaust at the workstation within the laminar flow area removes the potent compound from the airstream more quickly. The basic parts and other details concerning this particular approach are discussed in the following sections (Laminar Flow and Local or Point Exhaust).

E.   Laminar Flow Technology

This technology was developed primarily to protect the product or activity taking place within the laminar flow (or, more accurately, the unidirectional flow) environment. It does have application for containment when combined with the proper level of precautions. Analogous to the earlier discussion, laminar flow technology is made up of three basic parts: the physical structure, the internal environment, and monitoring systems.

1.   Physical Structure

The physical structure of a laminar flow system will include at least one open side where interactions take place. The integrity of the system on this open face is maintained by using a curtain of air moving at 90 ft./min. plus or minus 20%. Surfaces that can make up the sides of the structure should be smooth, with intersections of solid walls covered to increase the cleanability of the structure.

The surfaces covering the return air ducts should be removable to allow for cleaning. The design of the opening should enhance airflow and not create turbulence.

2.   Internal Environment

The internal environment is created by the directed movement of air to form an area separated from the external environment by a invisible air curtain. Particles becoming airborne inside this environment move toward the exhaust intake of the system and, as long as they do not encounter anything that disturbs the airflow, will move in a straight line affected only by the forces of gravity.

Placement of process equipment and support items (containers, for example) can cause the airflow patterns to deviate from design and need to be closely studied to assure their impact does not put personnel at risk. There are a number of computer models available to simulate the impact on airflow patterns by objects placed within the flow.

Personnel working with this mode of containment must be strongly supported by procedures, and in many cases visual aids, that identify proper positing of both the individuals and the materials that they will manipulate. Ergonomics is an important part of the evaluation used to establish these parameters.

Individuals working in the laminar flow environment will be required to place, at a minimum, their hands and arms into the containment zone. The system will require a means of decontamination of the protective clothing before the individual leaves the area. The preferred method is a liquid shower.

3. Monitoring Systems

Monitoring systems for laminar flow systems are focused at maintaining the airflows and condition of the filter systems. These systems should be alarmed to inform personnel of any system malfunction. Particle counters can have limited value in this particular instance because of an actual operation, such as weighing, could result in a very high count in the laminar flow area for a short period of time.

Outside the laminar flow area background counts can sometimes be too high to detect a meaningful level of difference in total particle count.

F. Local or Point Exhaust

Local or point exhausts are made up of the fan, filter, and duct system that removes particles from an area in which the particles become airborne. This technology has more failure than successes in industrial practice for two basic reasons: failure to understand and apply capture velocities for solid particles, and lack of detailed understanding of the operations being performed in the area where the extraction is taking place.

Many of the designs in the past have been based on the capture velocity of gases because this information is more readily available. To understand the capture velocity of particles, factors such as particle density, size, and shape are important. The face velocity of successful systems are often surprisingly high and result in larger and more expensive systems than expected.

The second factor resulting in failure to these types of system is that in many cases, the capture face of the system interferes with the operation. Because the unit is physically in the way, operators will bypass the system in order to do their jobs.

Good planning, emphasizing operational details, and an appreciation of the underlying physics involved in particle dynamics can yield a fine result with this technology. Lacking those elements, the system's chance of success is law.

IV. SUMMARY

Selecting the proper technology for containment of any individual operation, involving a potent compound from weighing to packaging, requires two levels of understanding and action. The first is to achieve a complete understanding and concensus of the activity or project. This should be expressed and communicated in a requirements document. This tool can be used for developing vendor specifications, tracking important elements of the project, and as a check list of things to do.

The second level of understanding is selecting the correct technology to provide an adequate level of containment. Table 1 defines the potential choices based on the acceptable exposure limit to be achieved. An economic analysis should be performed including all candidate technologies that meet the containment criteria. The total cost of each alternative needs to be included in this analysis.

8
Engineered Local Exhaust

Edwin A. Kleissler

Kleissler Company, Lakeland, Florida

I. LOCAL EXHAUST: ITS ROLE IN CONTAINMENT

Local exhaust is generally understood to mean developing air patterns at the source of emission that capture the maximum amount of dust emitted to the atmosphere from a manufacturing process or operation. The logic for using local exhaust is that containment is accomplished before particles have spread into a larger body of air. Control patterns are developed in the area of highest concentration of contaminant in contrast to developing an overall flow pattern in the general work area. Air volume is held to a minimum compared to other air-based methods; therefore, cost is held to a minimum. Local exhaust is a feasible technology for a wider range of applications because recent developments in the design of local exhaust control hoods have reduced the exposure levels that can be obtained. Local exhaust is an economical solution where it can achieve the desired exposure levels.

A. The Local Exhaust System

It is important to design local exhaust in the context of a system. A local exhaust system consists of a control exhaust hood(s), duct work, exhauster, filter separator(s), and controls. Of these components, the exhaust hood is the key element for achieving success. Without capturing the dust, the other components cannot lower exposure levels no matter how well designed. Therefore, we will focus on the exhaust hoods and only deal with other components in a selective way. There is much more reference material available to the design engineer on all other components.

129

II. A HISTORICAL LOOK AT LOCAL EXHAUST
HOOD DEVELOPMENT

Much of the basic work done in enunciating principles that govern good design of local exhaust was done in the 1930s and 1940s. Little was added as academics pursued new frontiers until recently with the use of new design tools. Today's engineer would do well by examining this early work and then, with modern tools such as the computer, take the care necessary for successful design.

The early development of the technology of local exhaust began around 1928 with academic attention being paid to the air patterns developed by hoods intended to capture dust particles. J. M. Dalla Valle of Georgia Tech and Theodore Hatch and Leslie Silverman of Harvard University did most of this work.

Dalla Valle recognized the rapid changes in velocity with distance that occurred in front of exhaust hoods. He saw that air came from all directions into the area of low pressure (1). Dalla Valle's main contributions were to analyze these patterns and, for the first time, establish equations for limited quantification of key factors. A given velocity contour was found to always have a velocity that was a fixed percentage of the velocity at the face of the hood which relates directly to airflow volume. Thus for a given hood shape the contour pattern and the streamlines were always the same regardless of the flow volume. For example, if a 30% contour line would have a velocity of 150 fpm with a volume of 500 cfm and the air volume were doubled to 1000 cfm, the 30% contour line would be at the same distance from the hood face and would have double the velocity or, 300 fpm. Thus a single diagram could be drawn for a particular hood shape. This also meant that exhaust hoods could be modeled for purposes of research.

Dalla Valle's means of arriving at velocity contour lines was experimental. He felt mathematical solutions were impossible. Even the simplest patterns were tedious and time consuming to plot. He designed a version of the Pitot tube that was accurate at low velocities but awkward to use properly. With this instrument, he took innumerable velocity readings on a number of hood shapes assuming no obstructions. These resulted in plots such as that in Figure 1. Note that Dalla Valle drew streamlines perpendicular to the contour lines. Doing so is a close enough approximation of the flow pattern for the engineer to gain a general visual image but is not strictly accurate. In fact, with more complex hood shapes than Dalla Valle worked with, there can be significant errors in assuming streamlines and velocity contour lines to be perpendicular.

Dalla Valle's next contribution was in deducing equations for centerline velocity. Plots of the experimentally obtained centerline velocities revealed shapes approaching hyperbolic functions and led to equations where velocity (as a percentage of face velocity) was a function of hood-face area and the distance from the hood face (inversely as the square of the distance).

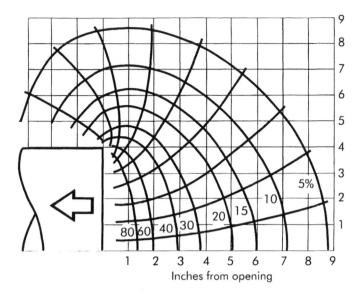

Figure 1 Streamlines as generated by Della Valle.

Della Valle's sketches of streamlines and velocity contours and his equation for centerline velocity have been used since to help engineers visualize flow into exhaust-opening configurations. They were the best available until recently when the advent of computers advanced the technology significantly.

At about the same time, Departments of Labor from major industrial states of the time, New Jersey, New York, Ohio, and Michigan, set up codes for local exhaust. These codes included air volumes for various metal and wood-working operations. Most people decided that everything was known about controlling dust from applications such as grinding wheels, table saws, etc. by using the air volumes described in state Labor Department codes and thus the "cookbook" approach began (2).

In 1946, the subject was new enough for Brandt and Steffy to publish a paper entitled "Energy Losses at Suction Hoods" (3). Energy losses for exhaust hoods were determined by testing 175 different hood configurations. Brandt and Steffy concentrated on obtaining a value for the coefficient of entry (Ce) for each hood shape. They determined the effect of different hood-design parameters including duct velocity, hood taper, face area to throat area ratio, use of flanges, flange width, large adjacent surfaces, and throat size. Their data contradicted much of what was used in practice at the time. W. C. L. Hemeon of the University of Pittsburgh provided a discussion of this paper. He widened the range of hood shapes and showed why length of taper made a difference in Ce. This was one of the first uses of research data in exhaust hood design. A designer today may wish to become familiar with their work.

Hemeon came along with the best analysis of the nature of the dispersion of dust and methods for estimating the air volumes required for control given a reasonable design of hood shape. Hemeon's book *Plant and Process Ventilation* (4) dealt with two particularly noteworthy ideas.

The first of Hemeon's ideas deals with the problem with determining the air volume to exhaust when dust is ejected from a process with force. This occurs, for example, when material falls and splashes as from a chute into a container. Hemeon used as an illustration of his solution to the problem a simple example, striking a chisel with a hammer. When struck by the hammer, the chisel causes dust particles to fly in all directions from the point of impact. He uses the term *null point* to describe the observed point where the velocity of the air carrying dust dies down to where the material moves with random air currents. He names the furthest null point from the hood face the X distance. Since random currents are usually between 50 and 75 ft./min. an exhaust capture velocity through the three-dimensional envelope drawn around the area at the X distance (Figure 2) should be in excess of 75 ft./min. to contain the discharged dust.

Another Hemeon concept was the effort to deal with elevated temperatures. A hot surface will generate an upward column of air. The volume of this column must be added to the control volume determined by using velocities through the contour envelope the engineer draws around the control zone.

The work of Drinker and Hatch in *Industrial Dust* (5) published in 1954 still remains a very useful reference. The bibliography contains 458 references, virtually everything written in that early age of development. It summarizes the work of Dalla Valle and Hemeon and covers important material on all aspects of dust and dust control. Particular attention is given to the quantification of air entrained by falling materials. The work of this early period deals with basics and is useful in designing for potent pharmaceutical compounds. Most pharmaceutical local-exhaust problems cannot be solved with "cookbook" approaches.

In the 1970s, a project was carried on based on the careful application to each individual operation of the principles of Dalla Valle and Hemeon. The improvement in exposure levels achieved compared to similar past systems attracted the attention of NIOSH. Exposure levels were not as low as can be achieved today with computer-assisted analysis of airflow patterns. However, the project did show that careful attention to the work of the early pioneers gave better results than the rule-of-thumb, cookbook-type methods to determine airflow volumes that were often used in the past.

The various editions of the *Industrial Ventilation* manual by the ACGIH (6) provide a wide range of useful information for the exhaust-system designer and are probably the best references available. This manual is most valuable for

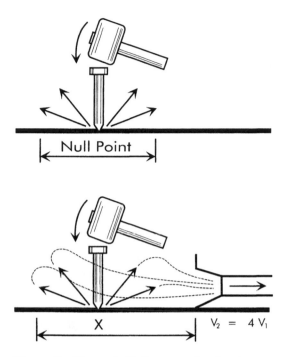

Figure 2 Hemeon's X distance and null point.

the sections on pressure drop in hoods. There is a good discussion of some of the basics ideas the engineer should keep these in mind when considering the use of specific recommendations. The applications illustrated are helpful in gaining ideas for pharmaceutical applications but specific pharmaceutical process operations are not covered. The applications covered use general rules for design such as airflow volume per area (cfm/ft^2) for surface areas or velocity (fpm) for openings. These rules of thumb work for those applications illustrated. It should be noted that most of the applications covered do not require exposure levels as low as those required for potent pharmaceutical compounds. Also, care must be taken when applying the approaches in the manual because of the many factors that can affect a general rule. Some of these influencing factors are described in this chapter.

At "Tissue World 97" (7) a paper was presented that discussed mathematical tools, now available to the engineer, that accurately model airflow patterns; provide velocity envelopes more accurate than those of Dalla Valle's methods; are useful with unusual shapes encountered in pharmaceutical processes; and provide a wealth of quantitative data as well as excellent color-coded visual plots.

The methods discussed include Computational Fluid Dynamics (CFD) and Flow Codes. The National Institute of Health (NIH), for one, is using a specialized type of CFD for tracking the spread of disease-carrying particles in a space with varying ventilation air patterns. These technologies bring the design process back to an engineered basis and greatly assist developing visualization in the mind's eye.

The recent use of analytical tools such as computer modeling and other methods that will be discussed in the following sections leads us to term those exhaust hoods that result from such use of development tools as ''modern'' exhaust hoods. Currently far from all local exhaust hoods in use and being put in use are modern. Data taken from installations that have not benefited from currently available design techniques will sell short the possibilities of local exhaust.

III. SOME HOMESPUN PHILOSOPHY

The engineer designing local-exhaust systems faces a myriad of different process situations, each with its own operating procedures and maintenance needs. The varying configurations and limitations require the engineer to visualize what might be the final shape of the exhaust hood in order to begin applying the various tools for good design. More importantly, the engineer should be able to visualize the flow patterns that will develop given the geometry of the situation. Air does not behave as the engineer might wish. The laws of physics determine its movement. Realistic visualization of the flow patterns that will develop will preclude the hopeful utilization of an inappropriate hood.

The difficulty in visualizing airflow patterns usually comes from a lack of appreciation of how velocities die off rapidly as distance from exhaust openings increase. Exhaust patterns are in stark contrast to air-supply sources that show a very high persistence in velocity levels contrasted with exhaust patterns. The difference in distance from the source for the same resulting velocity is shown approximately in Figure 3.

An engineer for a large supplier of high-velocity drying systems illustrates this point very well. The company's engineers thought they were capable of designing local exhaust but failed to achieve good results. The basic problem was that the engineers were used to dealing with air supply and not used to the behavior of air-exhaust patterns. The engineer telling this story said he illustrated the problem in a simple way. He held his hand several inches from his mouth and blew toward it. Then, with the same force, he inhaled and asked that others move their hand toward the mouth until they detected air motion. Of course, the hand has to almost touch the mouth. The air is approaching the mouth from all directions and therefore velocity patterns are weak.

Most of us have experienced cold drafts from air-conditioning supply registers in a restaurant and have requested a table where we can escape the discom-

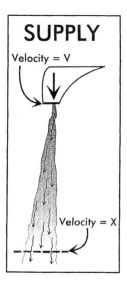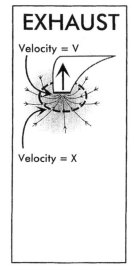

Figure 3 Relative supply and exhaust velocity fields.

fort. Who has sat close enough to an exhaust register to feel air currents? With distance, the velocities quickly drop off to those of random room currents.

An overhead kitchen hood looks like a very impressive capture device. However, upon reflection an engineer will realize the stove as a heat source causes a column of hot air to rise up to within the capture zone of the exhaust hood. The engineer will then be cautious of an approach to a process exhaust situation that has a heat source and will design with that in mind.

It is necessary for the local-exhaust design engineer to purposely develop the ability to visualize air patterns. This can be done through carefully making observations and by thoughtful study of the results of some of the technical approaches we will discuss. The first field analysis of a problem will not be productive without the air patterns and a possible design solution developing in the mind's eye.

IV. TYPICAL APPLICATIONS WHERE LOCAL EXHAUST HAS BEEN SUCCESSFULLY USED IN THE PHARMACEUTICAL INDUSTRY

Whether local exhaust can be considered successful for a particular situation in the applications that follow depends upon several factors. Most important is the allowable exposure level for the compound being used in the process. Also, the particular configuration of the process equipment being used may make it more or less likely to meet a given exposure level. Different materials have different degrees of dustiness. However, successful application of local exhaust has been made for the following applications albeit the criteria for success have differed for particular situations.

Aseptic Mill
Bench Scale
Bowl
Capsule Filling
Compression Hopper Charge
Dispensing
Drum Charge
Filling Machine Charge
Fitz Mill Charge
Fitz Mill Discharge
Fitz Mill Feed Tray
Floor Scale
Floor Sweep
Granulator Charge
Granulator Discharge
Mill
Mixing Kettle
Platform Scale
Reactor
Scoop from Drum
Scoop into Drum
Tableting
Tote Bin
Tray Drier Dump
Tumble Dryer Charge
Tumble Dryer Discharge
Turbo Sieve
V-Blender Charge
V-Blender Discharge
Weigh Pan

V. LOCAL-EXHAUST HOOD TYPES

Classifying hoods by application is most useful. Design shapes for the same application are often the same. This results from the designer relating the shape of the exhaust hood to the shape of the process equipment and recognizing that ergonomic considerations for the operator are similar. The methods of developing control air volumes are usually similar if not the same. The same types of parameters may be applicable. Results in terms of exposure levels achieved may be available and most useful.

Classifying exhaust hoods is helpful in the design process. For example, hoods of a similar shape may allow use of parameter-driven CAD design. Performance of similar-shaped hoods is more predictable.

Hoods for some applications are arc shaped, they encompass less than 180° of circles of various diameters. The included angles of flange and slot are probably the same for different radii. The shape of the connection to the exhaust duct is often similar. Complete ring hoods (360°) many have similar cross-section velocities and slot shapes to assure even exhaust patterns.

Classification as to whether the hoods are stationary; attached to equipment or mobile; or adjustable may suggest design features employed on previous designs.

VI. FACTORS AFFECTING DESIGN

The critical design factor is the exposure level to be achieved. The key question is whether the desired level can be achieved with local exhaust. If past experience assures that the desired level can be reached (see Exposure Levels Achieved in the following text) or, conversely, that it cannot be achieved, the process may be as simple as repeating a proven design already in a database or recommending an alternative approach (e.g., barrier technology).

Past designs may have been in the context of a different environment and a design may have to be modified due to conditions such as air currents in the space containing the process. If there is no prior experience to validate the use of an existing or modified design the engineer has to go back to the basics and deal with several other factors that influence design. Observations must be made and visualization of airflows developed.

If air is displaced from the process by mechanical means such as a piston, the ejected air may have a velocity vector greater than the control velocity vector that would have been used if the displacement mechanism was not present. Capsule filling is an operation that will cause forced ejection of air with dust.

Often a material flow is from a higher-level piece of equipment to a lower-level receiving piece. Falling material will entrain surrounding air and or induce airflow with the material. The volume of air so entrained will cause a flow of air containing particles to escape the receiving piece of process equipment. Loading a blender is an operation where material may come down a chute thereby entraining air as well as displacing the air in the blender. The exhaust hood must handle the volume of air entrained, the volume of air displaced from the receiving vessel, and the volume calculated to provide containment. The entrained volume and displaced volumes and the time periods of each must be estimated. Alternatively, the control envelope should be envisioned based on observation and the control velocity estimated based on experience to arrive at volume/time rate.

Another important design consideration is avoiding the pickup of excessive material. Excessive air velocities and the pattern of airflow may cause the pickup of material that should remain in the process. Flow across open spaces will dip into the open space (Figure 6). Air drawn across a drum will show this behavior unless the design of the flow minimizes the effect. The exhaust slot must be located sufficiently above the top of the drum, depending on drum diameter. Flow dipping into the drum may take airborne material that otherwise would settle out.

The size of the equipment is an important factor. For example, a large bowl will be more likely than a smaller one to have control air currents affected by room currents unless the design deals with the size factor. Baffles, partial covers, and using sufficient exhaust air may deal with the size factor.

A. Environments

The environment within which a local exhaust system is employed can affect its efficiency to a considerable degree. We will consider five environmental situations: 1) within a pressure controlled suite, 2) free-standing, 3) within a so-called laminar flow field (better described as unidirectional flow), 4) purposeful multidirectional flow, and 5) within a barrier-type enclosure such as a glove box.

1. It Is Usual for a Pharmaceutical Suite to Be Pressure Controlled

A pharmaceutical suite may be under negative pressure as an aid in preventing particles from escaping the suite or under positive pressure to prevent particles from entering the suite and potentially causing cross-contamination. The concern of the local-exhaust design engineer is whether the HVAC system will affect the local exhaust or vice versa.

The local-exhaust system can be starved for air if the HVAC system does not provide sufficient air beyond that to meet pressure requirements to replace the air removed by the local-exhaust system. One way to get around this problem is to return all the air from the local-exhaust system. However, this may not be good practice if the materials controlled are highly potent and/or hazardous to health. Precleaning to avoid loading replaceable secondary filters too quickly followed by one or more HEPA banks of filters may be acceptable but that is not always the case. The best way to deal with the issue is to coordinate the design of the HVAC system to deal with well-designed local-exhaust air volumes and potential for return to the space.

In the event of an existing facility where the HVAC system has been designed without allowing for a proper amount of containment exhaust air and the local-exhaust air cannot be returned, two possibilities remain. Alter the HVAC system to supply the needed additional air, or allow air to purposely leak into the room—that means it will be under negative pressure.

2. Using Local Exhaust in a Free-Standing Environment

A free-standing environment is one in which there is no protection against air currents such as in a booth or with protective walls or partitions. In this situation, the process may be exposed to random room currents or worse, for example significant air currents caused by HVAC or other mechanical systems. Thus, the process might be within a pressure-controlled suite or in a nonpressure-controlled location. It is critical to analyze the potential effects of mechanically induced air movement and to design the best hood shape possible. Well-designed baffles as extensions to the hood may be able to enhance performance. However, the basic hood design is of primary importance.

3. Local Exhaust Within a Unidirectional Environment

An environment such as a clean room or a booth will enhance the overall performance of local exhaust. To achieve the most from this combination, the local exhaust hood must be designed specific to the operation not merely an increase in airflow in the general area near the process.

4. A Multidirectional Airflow Environment

The most successful containment short of barrier or absolute isolation technologies is to combine careful shielding of the operation with local exhaust and auxiliary air currents that avoid buildup of contaminants. The auxiliary airflows will be of more than one direction to take into account the operator's position, the ergonomics of the operation, and the need for containment within an enclosed area. The flows must be carefully coordinated (Figure 4).

5. Barrier Environments

The barrier environment is discussed in the following text under Future Needs and Applications of Local Exhaust.

B. Finish and Cleanability

It is generally expected that the dust-control components visible in a suite will be finished to the same degree as process equipment. This is usually accomplished by using stainless steel with a Number 4 finish, although material and finish

Figure 4   Multidirectional airflow environment.

requirements should be confirmed with Quality Assurance or process-engineering representation for the specific application.

Exhaust hoods used in local exhaust systems are in close proximity to the process operation. Being able to clean the hoods is, therefore, an important design consideration. External cleaning is always important and usually it is deemed necessary to be able to clean the interior of the hood. Cleanability is a function of material, fabrication methods, and access.

A pharmaceutical hood is normally fabricated of stainless steel. Ideally a stainless sheet with two sides polished is used. A finished hood requires dealing with polishing of the welds and the overall surface. Two welding methods may be employed metal inert gas (MIG) or tungsten inert gas (TIG). MIG is a wire-fed welding process that means material is added to the weld line. The time to grind and polish the MIG welds offsets the savings from the faster speed of MIG welding. The use of TIG is much preferred. TIG is essentially a fusion process with only occasional use of hand-fed wire. TIG uses less heat and consequently results in less distortion. With proper polishing techniques it is usually difficult to see the weld. Also, inside corners are tight.

If inside TIG is required, the time for welding and polishing is considerably greater. Also, the design must allow a carefully planed sequence of assembling sections of the hood to give access for welding. Sometimes a more expensive take-apart design is necessary.

Access to the inside of a hood is simplified by having an adequate slot size so that a hand-held wiping cloth can be inserted. Concern over large slot sizes is misplaced. A high-velocity slot opening does not improve the exhaust air pattern beyond a very short distance from the slot.

Access doors must be provided to clean areas beyond those that can be reached from the slot. At times, the hood must be taken apart. An example is the claw-type hood for a capsule-filling machine illustrated in Figure 5 and further discussed in the Some Special Designs section. The "claws" not only swing out but lift out by the use of special hinges. Remote cleaning is then possible.

C. Ergonomics

The local-exhaust hood must remain in the location for which it was designed to function successfully. If the operator is inhibited in performing the required operation, chances are the exhaust hood will be moved and its effectiveness diminished or eliminated.

If the exhaust hood is designed to be moved to allow access for the operator or to place the hood properly, consideration must be given to the weight. This is true of vertically adjustable hoods such as those used in dispensing. Linear bearings may be used. Alternately, the hood may be stationary and the process element may be moved to the hood. Again, using dispensing as an example a lift may be used to move drums vertically to the designated height relative to the exhaust slot.

When exhaust hoods are permanently attached to a moving piece of equipment, such as one that rotates, operators are relieved of the burden of locating the hood properly.

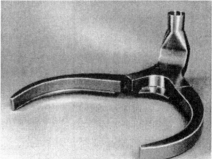

Figure 5 Capsule filling hood disassembly.

VII. DESIGN TOOLS AND METHODS FOR LOCAL-EXHAUST HOODS

Several design methods are available to the design engineer. Often these methods are complementary and should be used in combination rather than in isolation. Building a prototype and testing it is, occasionally, a necessary step in design. See the Testing section.

A. Using an Exhaust-Hood Database

Selecting exhaust-hood designs from a database is an economical process. Some thoughts on maintaining a database include the following.

The primary approach is to have a system for calling up past hood designs that are similar.

1) Similar in application
2) Similar in shape
3) Similar in the method of computing air volume for control
4) Usually close in cost

The database is also a means of finding particular projects on which such hoods were used.

1) When we cannot recall the location in which the particular hood was used
2) To see which past project would provide us with the information we seek
3) To locate a hood drawing

Use of a relational database to be able to combine parameters in a search.

1) Hood type and application
2) Hood type and size

B. Parameterized Design

It is reasonable to assume that the drawing of hoods will be done on a CADD system. The potential for real design with CADD is in the use of a parameterized version. Certain relatively standard-shaped hoods can have a drawing that will adjust to changes in key parameters according to design rules that are imbedded in the code. The parameters and calculations for the design rules can be on a spreadsheet that is linked to the CADD drawing.

C. Field Observations

Working with a hood database or with parameterized design assumes that good designs already exist. The range of operations, environments, and influencing design conditions is so great that the design process often has to start at the very beginning. It can not be over emphasized that the observations of room air currents from HVAC or other processes or air movers must be accounted for in deciding hood shape, baffles, and control velocities.

If the design engineer has not worked with local exhaust on a particular operation, the first and most vital part of the design process is a set of observations in the field. If the particular operation is not running or it is a design for a plant yet to be built, a comparable operation should be sought out for purposes of observation. In the latter case, the influencing factors mentioned in a prior section would have to be taken into account possibly with computer modeling. Integration with HVAC in the original process/facility design is important.

If a careful observation of the dust emitted under force is made, the engineer can use Hemeon's X-distance approach to help design the control envelope. Determining direction of the emitted airstream and the distance to where the dust is seen to move with random room air currents provides one level of control envelope to be established. Temperature effects must be noted. Note other conditions that will affect calculations for example altitude. Air density decreases with increasing altitude. It is better practice to design for a movement of a mass of air rather than a volume of air when altitude is a factor (usually over 2,000 ft.). It is a mass of air molecules that causes containment.

D. Fluid Flow Photos

The low-pressure drops encountered in local-exhaust systems allow the flow to be regarded as incompressible. Thus the design engineer can learn about flow patterns from published photos of incompressible fluid flow and, for that matter, from watching water flow in nature such as a mountain stream with changing velocities and vortices. Learning to envision flow is important in the design process. The engineer designing the domed arc hood described previously envisioned a shape in his "mind's eye" before starting the formal drawing process. This comes about in an effective way by having a "feel" for how fluids, including air, flow. Examples of observing flow over a reservoir are the photos in Figure 6. Note the dip in the reservoir changing when the ratio of the width of the reservoir to the depth changes from 2 to 3 (8).

E. Enhanced Visualization of Actual Flow

When testing a prototype hood it is important to be able to see the patterns developed. Smoke tubes will emit a point source stream that is useful for concentrating

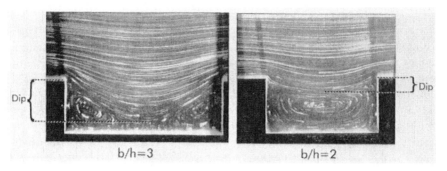

Figure 6 Flow over a reservoir.

on what is going on in a specific limited area (Figure 7). Fog generators give overall flow patterns. Fog will also be valuable in seeing the room air currents as well as the detailed flow in the vicinity of the hood.

Videotaping the smoke and fog patterns is helpful for further study and also for making comparisons of different designs.

F. In the Mind's Eye

Eugene S. Ferguson (9) maintains that good engineering is still as much a matter of intuition and nonverbal thinking as of equations and computation. ''The mind's eye is a well-developed organ that not only reviews the contents of a visual memory but also forms such new or modified images as the mind's thoughts

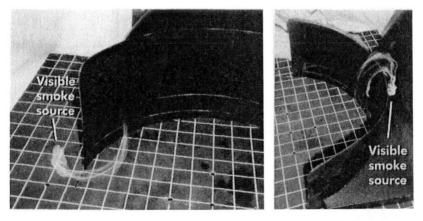

Figure 7 Depicting airflow near a flange.

require. As one thinks about a machine, reasoning through successive steps in a dynamic process, one can turn it over in one's mind. The engineering designer who brings elements together in new combinations is able to assemble and manipulate in his or her mind devices that as yet do not exist.'' All the design methods we mention only help the engineer form in the mind's eye the overall design of the local-exhaust hood. I believe Ferguson would regard the design of local-exhaust hoods as a low-level invention. The image that is conceived in the engineer's mind is the key element in good design. A design will not spring forth from a computer model or from a CAD drawing unless these are guided by the engineer's sense of a direction toward a final result.

G. Computer Modeling

Major changes have occurred since Della Valle lamented the lack of mathematical means to determine exhaust-hood velocity patterns. The theory of potential flow has been further developed; simplifying assumptions for the Navier Stokes equations have been developed and their usefulness validated; and most significantly the computer has become a day-to-day tool.

Computer modeling is a design tool that helps in deciding whether flow patterns can be improved. Physical modeling is difficult, and can be expensive and time consuming. It is far better to have a tool to depict flow that would permit changes in geometry and air volumes to allow considering alternative designs for increased effectiveness and/or lower cost.

Types of computer modeling include full-scale Computational Fluid Dynamics (CFD), modified CFD developed for ventilation purposes, and Flow Codes adapted for exhaust-hood analysis.

Full-scale CFD, because of its complexity, does not easily allow constant changing of shapes to work with. It takes appreciable time to develop models. CFD requires specialized staffing with consequent cost. Its power may, however, be justified for a critical problem or a repetitive situation where the cost can be justified.

Specialized ventilation CFD is helpful when dealing with an air movement in an entire suite but not with certain shapes such as those used with exhaust hoods. While it is easier to work with than full-scale CFD, occasional use may be best dealt with by using the consulting group of a company that offers such a program on a contract basis.

Flow Codes depict flow patterns accurately, provided they are used in situations where an assumption of potential flow can be made. That is, the flow is incompressible, inviscid, and irrotational. Flow Codes are derived from Panel Code technology that has been used in the aircraft industry since the advent of the modern computer in the 1950s and 1960s. Panel Codes and hence Flow Codes, using the same mathematics, divide surfaces into panels much like the process of finite element analysis for computing stresses and then computing surface speeds. From these, the flow at all points in the relevant field can be computed. The chief advantages of Flow Codes over CFD are a more simple user interface, quick changes in geometry or air volumes, and rapid run times with desktop computers. The disadvantages are that Flow Codes cannot correctly model free jets, recirculation zones, transport of contaminants, or thermal effects.

H. A Case Study to Illustrate Design Techniques Used to Develop a Modern Local-Exhaust Hood: a Domed, Mobile, Adjustable, Arc Hood

We will illustrate the use of different design techniques through the design process that was used for a proprietary hood design in which the author participated. The result of the design process is a hood with unusual baffling and new parameters for slot size and location. We will refer to this hood as a domed arc hood.

The so-called fishtail hood is shown in manuals such as the *Industrial Ventilation* manual. It is a semicircular-shaped hood with a slot over most of its face and a small flange rather than a raw edge around the slot to increase the coefficient of entry. The effect of the lower coefficient is a greater volume for a given pressure drop. More importantly, the presence of a flange also improves the flow pattern in front of the hood.

For a long period of time, it was accepted that this type of exhaust hood, indeed most local-exhaust hoods, had an efficiency limited to achieving short-interval exposure levels in the workers breathing zone and in the area surrounding the worker in the milligram per cubic meter (mg/m^3) range; often mentioned were values of 1 to 5 mg/m^3. The challenge was whether the efficiency of this type of local-exhaust hood could be improved using careful engineering analysis. A range of computational methods and visual test techniques were employed followed by careful testing of the results.

At first, the traditional fishtail design hood was mounted on a flat surface with a painted grid and series of holes at critical locations on the grid. By blowing smoke through the holes, one at a time, it was possible to concentrate observations of flow patterns from the various points. Most striking was how the smoke flowed around the flange at high velocity (see Figure 7). This points up that the flow into an exhaust opening comes from all directions. As with all subsequent visual patterns developed in various ways described in the following text, the results were photographed and captured on video for further study and for purpose of comparison.

The same hood was then placed at a drum in order to make observations in a typical application. For these observations a fog generator rather than smoke was used so that a continuing large amount of fog could be generated. Fog was discharged within the drum and from various locations on the periphery of the drum. This procedure was videotaped. It was clear that the motion around the edges of the hood flange was turbulent and that wisps of fog were escaping. The degree to which the fog released on the far side of the drum from the exhaust slot dipped below the drum surface was also of concern since exhaust air dipping below the drum surface might take material from production by entraining particles that would have settled back into the drum. Figure 8 shows a domed arc hood with fog generated in the drum.

In addition to using fog as a dusty material, finely milled cornstarch was scooped into and out of the drum to simulate dispensing a common pharmaceutical operation. Mixed with the cornstarch were fluorescent particles. By use of an ultraviolet light it was possible to see where escaping particles went.

It was obvious that the traditional design of such a hood, while useful for working with some innocuous materials, was inadequate for the more demanding performance required in pharmaceutical operations.

After numerous tests, a number of design parameters were being discussed and at the same time an image of what might be a preferable design began to form in the mind's eye. Parameters included the degree of wrap around the drum; the included angle of the exhaust slot; the shape and size of the exhaust slot; the volume to be exhausted; control velocities at critical points; the size and shape

Figure 8 Flow of fog from a drum into a domed arc hood.

of flanges; the distance of the hood from the drum; and the distance of the exhaust slot above the lip of the drum. The envisioned image was of a complex shape. It was decided it would be time consuming and expensive to test combinations of the design parameters by building prototypes and running tests with fog and so on. Instead the decision to use a computer model was made. By using Flow Codes that were adopted from the aerospace field, changes in parameters could be easily made and the flow patterns obtained in minutes (see Figure 9). In addition, a limited version of full-scale CFD was used to model and detect vortices that would tend to trap particles and, when disturbed, release particles to the atmosphere.

Finally, the prototypes for the most likely successful designs were fabricated. These designs all had curved side-inlet flanges and a partial dome over the top toward the slot. Further fog tests were made and refinements incorporated.

Exposure-level tests for typical transfer operations were conducted. Test techniques used were by tracer gas and by weight (using an air pump and filters as per NIOSH procedures). These showed an order of magnitude reduction in levels compared to the fishtail hood.

Mechanical design was next required to make the hood adjustable vertically and to be mobile, thus enabling it to be used in various locations in the facility and with various height drums. The result is shown in Figure 10.

Computer Flow Code

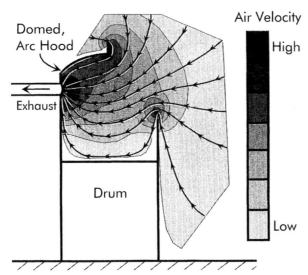

Figure 9 Computer-generated airflow pattern into a domed arc hood.

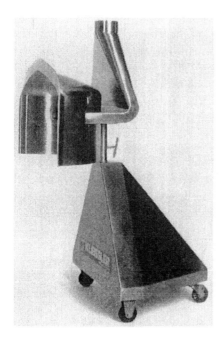

Figure 10   Adjustable mobile dome hood.

The same types of design techniques have been used on a wide range of local-exhaust hood designs.

VIII.   SOME SPECIAL DESIGNS

There seems to be a limitless number of configurations needed to satisfy the needs generated by different operational process configurations and standard operating procedures in the pharmaceutical industry. The following examples of hoods are representative not only of the range of designs but also illustrate the application of good design principles.

An example of a hood that provides excellent control and also prevents undesirable pick-up of material being processed is the air seal–type design. We show a feed hopper over a Fitzmill in Figure 11. The exhaust hood covers all four sides of the space above the mill. Airflow to the exhaust hood will not allow dust to escape upward and will only take away product that would have escaped to atmosphere in the space around the hopper. An exhaust hood with a more direct access to the stream of falling material would remove more material than is necessary or desired.

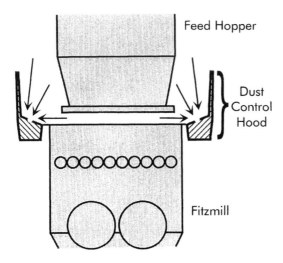

Figure 11   Control pattern for a feed hopper over a fitzmill.

The claw-shaped hood described in a previous section on Finish and Cleanability is designed not only for cleaning but also has removable arms in order to install it on the capsule machine. Its shape is an exact match to the platen on the machine so as to generate flow across the platen to keep the area clean as well as to develop a flow pattern to keep particles from escaping into the atmosphere.

The interior exhaust opening in the removable reactor hood (Figure 12) keeps the reactor under slight negative pressure while loading the reactor and is connected by duct work (the lower connection) to a scrubber to handle the vapors. The exhaust slot above the opening is to deal with dust that is generated when loading the reactor.

The tray dump hood (Figure 13) allows a tray to be inserted into the slot. An exterior handle causes the tray to turn over releasing material to the container below. Exhaust air controls the movement of air into the chamber but in a pattern that does not cross the falling material, thereby avoiding removal of excessive production material.

The V-Blender hood (Figure 14) is attached to the V-Blender and rotates with it. It is always in the correct position when in use. A flexible exhaust duct is attached by a quick disconnect fitting when in use. An alternative design is the retracting V-Blender hood (Figure 15) that uses an air cylinder to move the hood into the correct position when charging and out of the way when blending.

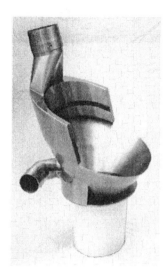

Figure 12   Reactor hood with separate dust and vapor removal.

Figure 13   Tray dump hood.

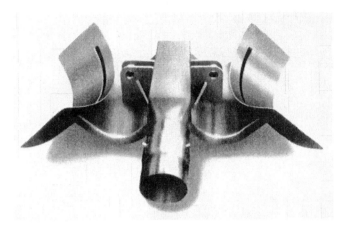

Figure 14   V-Blender hood that attaches to blender.

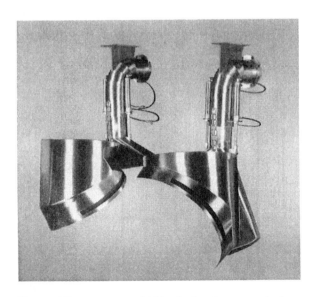

Figure 15   Retractable V-blender hood, not attached.

IX. TESTING

Types of testing include testing a prototype hood for flow patterns, testing for personnel exposure levels and testing for exposure levels for engineering design purposes.

An example of testing a prototype hood is the sifter hood shown in Figure 16. The prototype test seemed to be necessary because of the operational requirements. For one, operators actually work the material through the sifter with their hands. Secondly, the operator has to have visibility. A sleeve was extended from the hopper down past the ring-type exhaust hood. The question was whether the air pattern would capture the dust with the particular configuration. An effective design resulted.

Testing for exposure level is a subject in itself, and the following is worth noting: In testing local-exhaust hoods in a workstation that has special directed-air patterns, the expectation is that the levels can be in the low microgram ranges. However, the variations that occur in testing by weight will not allow good measurements at low microgram levels. These variations come from handling the filters and the changes in filter weight from vapor pressure. Testing with a surrogate that has a profile that allows the use of High Pressure Liquid Chromatography (HPLC) is a solution to that problem. HPLC measures only the material being tested, not other dust that may be in the atmosphere. It is a very accurate measure without the problems associated with the weighing method and hence valuable when the levels are very low.

Testing personnel is accomplished by attaching filter inlets on the operator's collar with tubing running to a pump worn on the belt. Testing for engineering purposes requires a number of readings in the area around the exhaust

Figure 16 Prototype hood for testing a sifter hood design.

hoods to see where particles tend to escape. This latter approach can be supplemented by using fluorescent particles mixed with the test material and seeing with an ultraviolet light where the particles are that did not enter the hood.

X. EXPOSURE LEVELS ACHIEVED

It is common practice to look at the absolute level that is achieved with a given approach to containment. More important for a given application is to have data that provides a measure of the reduction that can be achieved in order to make a judgment as to the applicability of the method. Levels on an operation can be measured; the percentage reduction achieved on a completely similar application applied; and the result compared to the level necessary to achieve. As discussed elsewhere, the final result is a function of the material itself, its dustiness as well as the design.

However, it is helpful to know what magnitude of exposure levels can be achieved with any method including local exhaust. Three carefully tested examples follow.

Tests conducted on the domed arc hoods whose development was described earlier resulted in levels of approximately 100 $\mu g/m^3$ (10). The operation tested was scooping milled cornstarch, a very dusty material, from one drum to another, each drum having a domed hood. Personal sampler tests were conducted using a weight procedure in accordance with NIOSH. The results are for the life of the task not time-weighted averages. These results are probably conservative. The task was run for time periods longer than those encountered in actual pharmaceutical operations allowing background levels to build up.

The tests conducted on several operations in the case discussed in the Taking Advantage of Local Exhaust section gave results that were in the range of 150 $\mu g/m^3$. Tests on some operations were as low as 30 $\mu g/m^3$ (10). All results were for the life of the task. Personal sampler tests were conducted using a weight procedure in accordance with NIOSH. These tests were also run for long time periods allowing background levels to build up.

A major pharmaceutical company tested a proprietary multidirectional airflow workstation with local exhaust for dispensing. Personal samplers and background samplers were used on repeated trials using their Standard Operating Procedures. Test results were obtained using HPLC. The results were in single-digit micrograms per cubic meter (10). These results were for the life of the task. Both personal sampling and area sampling produced these results.

Obviously, the design of hood employed, the surrounding environment, operating procedures, and the material being processed affect exposure levels. However, the range of situations cited indicate that with modern design hoods, results in the low microgram level can now be obtained rather than in the level of 1 to 5 mg often quoted only a few years ago.

XI. SYSTEM CONSIDERATIONS

We have focused on the exhaust hood, the design of which is the key to successful engineered local exhaust. However, the hood is part of a system; all parts of a system must function well for the system to achieve desired results. We will now pay attention to some of the important ideas for good total system design.

The simplest local-exhaust system is a single source system, one with only one local-exhaust hood. The multisource system inherently has many complications, some of which are discussed separately in the following text. The following system design considerations are those for single-source or multisource systems as well.

A. Duct Design

Good duct design allows only minimum use of flexible exhaust duct. When used, it should be a sanitary type and should have minimum bends to maintain low-pressure drop. Since dust may build up a charge when flowing through duct, grounding should be provided to the wire frame in flexible duct.

Conveying velocities should be maintained with a safety margin. Velocities of 3800 to 4200 fpm are usually a good balance between initial cost and running energy cost but can vary depending on a variable range of energy costs throughout different regions globally.

It should be noted that higher velocities might be used at higher altitudes without an energy penalty because duct friction is lower.

Good duct design includes the use of wide radius elbows; flanged and sealed connections; and grounding to avoid the buildup of static electricity charges. This is especially important if the specific powder has a low minimum ignition energy (MIE).

B. Dust Collector (Separator)

There is a considerable body of technology on dust collectors. We will touch on some considerations that are useful for pharmaceutical applications and will discuss the most-used filter type. The type of collector should depend on the need.

- A collector with intermittent cleaning is sometimes adequate and is cost effective if the dust control system can be shut down at intervals of four to eight hours. This is sometimes the case in short-run pharmaceutical processes.
- A collector with continuous automatic cleaning should be used if the system cannot be shut down at intervals for cleaning or when dust loadings are particularly heavy causing the collector pressure drop to increase to the point of affecting system performance. Cartridge filter collectors are continuous automatic. Filter cloth separators may also be continuous automatic.

Bag-in Bag-out is often a requirement for pharmaceutical systems for protection of the workers when changing filter media. A filter deemed faulty is pulled out into a plastic sleeve that is attached by an airtight connection (Bag-out). New filters are inserted in a reverse procedure (Bag-in). Both procedures protect the worker and the atmosphere from the dust in the separator. The cartridge-type collector is well suited for this operation.

Proper provisions should be taken for conditions where humidity is present, VOCs are in the airstream, or corrosive gases are present.

Air-to-cloth ratio is a key factor in making a filter-type dust-separator selection. Particle size is a determinant. Uniformly small particles cause a tight filter cake and may raise the collector pressure drop to unacceptable levels. Having all submicron particles can actually blind a filter unless the ratio is very low (less than 1:1 for intermittent collectors). The collector manufacturer should be able to make a recommendation for air-to-cloth ratio. In general the air-to-cloth ratio for a cartridge-type filter will be half that of an intermittent collector and the intermittent collector ratio will be half that of a continuous automatic collector.

Hazardous dusts may also require explosion vents or an explosion suppression system. Sprinklers should always be provided. Dust should not be allowed to accumulate in collector hoppers in order to minimize the material present in the event of a system fire or explosion. Rotary valves to a sealed container are preferable.

Redundant filtration is often a necessary precaution for recirculation systems or for once-through systems depending on outdoor environmental criteria. HEPA filters are the common second filters and on occasion these may be in series. HEPA filters are available in Bag-in Bag-out (safe change) configurations. Dr. Melvin First provides excellent coverage of HEPA filters in a chapter in the book *Treatment of Gaseous Effluents at Nuclear Facilities* (11).

A decision must be made whether to design a central system with one dust collector or a system with multiple dispersed collectors. The decision must be based on the circumstances of the individual system. An important influencing factor can be achieving a balanced system.

C. Exhausters (Blowers, Fans)

A large body of information is available on exhausters. For pharmaceutical applications the following features are recommended:

Type C spark proofing at a minimum
Shaft seals
Access doors
Long-life bearings

Careful consideration should be given to noise level. Attention should be given to the noise carried through the ducts since most pharmaceutical operations are conducted in an environment that is quieter than many industrial applications. Noise generated through the housing can be dealt with by using sound enclosures (provide sufficient ventilation for the motor) or by wrapping the body of the fan with appropriate material.

D. Multisource Systems

Multisource systems are more common and have the complication of seeing that the proper air is exhausted from each exhaust hood. This is known as balancing the system and is often spoken of without due consideration of the difficulties in achieving it. These difficulties stem from attempts to minimize the overall system size and, therefore, design a dampered system; from ignoring the HVAC implications for positive or negative pressure suites; from allowing operator control of dampering; and from a failure to design the overall system configuration with proper accounting for balancing.

1. Balancing with Duct Diameter

Duct diameters are selected when the system is designed. If the pressure drop from all the exhaust hoods is equal the system will be in balance. This method of balancing assures that the duct route with greatest resistance maintains the design transport velocity. Duct diameters may have to be reduced in other branches to increase velocity and, therefore, pressure drop to where it equals the pressure drop in the duct run that had the greatest resistance with normal transport velocity. As a result, there is little chance for dust to settle in the ducts. Another advantage is that very accurate calculations are possible. It is the least expensive alternative and provides a system the operators can't throw out of balance the way they can with a dampered system. On the other hand, it is not a flexible system. Changes in air volumes at a given point cannot be made if field experience indicates the design volume is not what is needed. Also, calculations involving the small diameter ducts normally found in pharmaceutical operations may call for fractional sizes that are not available commercially thereby limiting the application of duct diameter balancing.

2. Dampers (Blast Gates)

Dampers are the most common method of balancing. They allow for field balancing and the system may be designed to allow some change in air volume particularly where the designer desires a contingency adjustment. However, calculations require empirical data that is dependent on damper design and adjustment is very sensitive. Change doesn't begin to occur until the damper extends well into the duct. Each small adjustment then quickly causes a change in air volume. The design should minimize the possibility of particle accumulation on the dampers. And while it is sometimes desirable to allow operators to change volume this can give poor control performance and also throw the overall system out of balance, particularly when going to complete shutoff.

3. Orifice Plates

Orifice plates may be inserted between a pair of flanges. The hole diameter necessary to achieve a given pressure drop, given branch design velocity, can be readily calculated. Orifice plates can be changed giving some flexibility to balancing the system. Operators cannot make changes easily, thus safeguarding the balance of the system. Particles can accumulate, however, and field balancing is not as easy as with dampers.

4. Venturi Sections

More sophisticated than orifice plates, but more costly, are venturi sections. A venturi consists of a tapered contraction, a throat section, and a tapered enlargement. The pressure drop across a venturi can be calculated accurately and there is no place for particles to accumulate. Operators cannot make changes. However, field changes require removing a venturi section and replacing it with another at some cost.

5. Self-Adjusting Venturi

Commercial self-adjusting venturis are available. They are adjustable and very sensitive to a change in airflow. The result is constant volume at the exhaust hood. The effect of dust on their operation is now known. It would be wise to install them in a horizontal section of duct to minimize any chance of buildup. A new design, now only in the prototype stage, claims to handle dusty air. The developers promise accurate performance and output signals for monitoring.

Usually a combination of duct diameter for approximate balancing and one of the other methods in combination is seen to be the best approach.

E. Controls

Control of system air volumes exhausted in a balanced system is critical to be certain that proper air patterns are being developed at each of the local exhaust hoods. Control of the system may be embodied in a PLC (Programmable Logic Controller), or connected electrically to a signal system, or depend on manual observations.

The PLC approach may be used if sequencing with HVAC or automated setting of dampers is desired.

Alternatively, a signal light system may be used to indicate that a particular line may be put in use (for example by opening a damper) without upsetting the system by having too many branches of a dampered system in use at a given time. Signals could also be operated if pressure drops change beyond a preset tolerance as measured by pressure gauges.

Manual observations of pressure gauges are often used because the operations are usually quite stable. The dust collector and HEPA filters are the key system components to be checked. The primary dust collector and each stage of HEPA filters should have separate instrumentation. A reading of increased pressure drop across the dust separator will alert the user to possible clogging of the dust collector filters that in turn will reduce system air volume. A pressure-drop reading beyond the recommended level on HEPA filters will indicate they are due for replacement. Similarly, a reading of pressure drop across the exhauster will indicate if a problem exists somewhere in the system including, of course, the dust collector or HEPAs. Inexpensive instruments are available for such readings.

While discussing balancing we mentioned the use of a self-adjusting venturi to automatically compensate for changes in a system's static pressure in order to assure maintaining a constant airflow in a single exhaust line. A bank of these units to handle a large air volume can be used to maintain volume control of a complete system.

The cleaning cycle of continuous automatic dust collectors commonly known as pulse jets is controlled by a timer that cycles a set of solenoid valves that open diaphragm valves. Each diaphragm valve allows a pulse of compressed air to be released from a pipe with openings over each filter tube in a given row to displace the dust accumulated on the outside of a round filter tube.

The cleaning of the filter tubes on an intermittent dust collector is accomplished by shaking the filter tubes to knock down the dust that accumulates on the inside of the round or panel-type cloth filters. A timer that activates the shake cycle when the system is shut down commonly controls the cleaning. It normally allows a short time period of several seconds for the fan to slow down before a shake period of one to two minutes is activated.

It is important that collected dust does not escape the filter system, particularly if the air is to be returned to the space. A backup filter, usually a HEPA, will load quickly if a leak develops in the primary dust collector. The HEPA's pressure drop will increase as shown by the pressure differential gauges. A dust collector can be tested for leaks by introducing fluorescent particles whether through holes in filter media or from seals.

Intermittent tests of filtered air may be conducted using iso-kinetic sampling and weighing a sampler filter. Continuous emission monitors are available.

The types and characteristics should be carefully evaluated for the low-level dust loadings common in pharmaceutical systems. A discussion of these methods is beyond the scope of this chapter.

XII. A CASE STUDY USING MODERN LOCAL EXHAUST HOODS

This case study is useful for a few important reasons. It illustrates the difference in efficiency between the use of old, more traditional exhaust-hood designs and designs arrived at by careful use of the development methods described in the section Design Tools and Methods for Local Exhaust Hoods that we have called modern hoods. The measured exposure levels achieved are less than those mentioned in the literature and in several papers presented in various pharmaceutical symposia and related courses in the past. Probably most important, as a result the project was carried out in such a way that predictions of efficiency are available for several of the owner's process steps.

The project was carried out for a major pharmaceutical firm with close collaboration among the owner's certified industrial hygienists, operating managers, and the engineering manufacturing company that performed under a total design, fabrication, installation, and joint-testing contract. There was an existing system in place for certain of the process points where containment was desired. The owner defined a maximum exposure level for the potent component of the product being processed.

Initially, the owner conducted personal sampling in the areas of concern for the project. Several operations were involved including blender charging and emptying; milling; and weighing. An existing dust-control system had local-exhaust hoods in place. A new system involved replacing the existing system's local-exhaust hoods with new hoods using currently available design practice. Tests were conducted after the system was up and running. The test procedures duplicated those used before the new system was in place. Some direct comparisons were available.

Containment efficiencies where a direct comparison was clearly available improved from the original "traditional" design by an order of magnitude—a reduction of over 90%.

In future applications, the owner will take personal sampler and fixed-point readings. On a given piece of equipment used in the same way with the same standard operating procedure it is quite possible the exposure levels will be different from those obtained in this case study. The material handled may have different characteristics such as particle shape and density. It may be more or less dusty. Therefore, the exposure levels achieved in this case study should not be used in an absolute sense.

This case is discussed in the Exposure Levels Achieved section. However, the percent reduction at individual process points in this case may be applied to the readings in another situation with different materials being processed. In this manner, probable levels can be calculated that will be achieved with the same type local-exhaust design. A decision can then be made as to whether local exhaust will meet the requirement for the material to be processed.

XIII. TAKING ADVANTAGE OF LOCAL EXHAUST TO ACHIEVE PRODUCT ACCOUNTABILITY

It is often important in the pharmaceutical industry to account for all the material brought into a process stage, such as tableting. In other words, an overall material balance is sought. Some of the material escapes the process and gets into the plant atmosphere. Some of this material may be entrained in large air volumes such as in the HVAC system, large booths, and in clean rooms. It is very difficult to measure the quantity of material brought to large HEPA filters. However, with local exhaust the air volumes are small and the percentage of material that gets into the air system is higher than in general exhaust systems (rather than settling on surfaces). The small volume allows for precleaning the airstream with super-high efficiency cyclones (Figure 17). Not only does this allow weighing the material collected by the cyclones but also it prolongs HEPA life by a multiple of 50 or more.

XIV. FUTURE NEEDS AND APPLICATIONS OF LOCAL EXHAUST

There is a need for automated monitoring and control systems to assure performance levels will be maintained.

Consideration should be given in the future to the use of additional air currents to supplement the exhaust air pattern. This in its simpler form is known as Push-Pull. A supply of air is directed to the exhaust pattern to assist in control of dust particles. The supply air is a form of jet. There is little information on subsonic jets that can be used in solving pharmaceutical local-exhaust problems. Empirical data must be developed to supplement the theory and to be able to use numerical methods on computers.

Largely under development, rather than in actual use, are miniature local-exhaust sytems within barrier environments. The purpose of these systems is to minimize the build-up of material within glove boxes or other barrier-type environment to better facilitate maintenance or process changes by lowering dust concentrations for these situations.

Figure 17 Super-high efficiency cyclone for product accountability.

The author was involved with the design of miniature systems within lab hoods including a five-hood system measuring less than 18 in. in length for the first transistor line at Bell Labs. Within this concept there are a number of design issues unique to the size not the least of which is an exhauster handling the small air requirements containing particles and developing the necessary pressure drop.

For some situations a complex hood shape is desirable; alternative materials and methods of construction should be explored. For a standard application on a particular piece of process equipment the possibility of casting a shaped hood is appealing. Criteria for the material would include conductivity to avoid building up static charges and cleanability.

XV. SUMMARY

Local-exhaust systems today provide reduction in exposure levels that are usually greater than what has been reported in most pharmaceutical symposia. In fact, local exhaust is a key factor in air-system solutions that bring exposure levels to low microgram per cubic meter levels. The engineer who undertakes local-exhaust system design will do well to concentrate on the local-exhaust hood while also understanding that all components of the system must be designed well. Concentrating on local-exhaust hood design entails using theory that is often found in papers published in the past; appreciating the rapid decay of velocity with distance from hood exhaust openings; developing the ability to envision flow patterns; and using a range of tools that are now available including parameterized CADD and computer modeling. Integration with HVAC is often critical. It is important to study the ergonomics at the operator interface to assure acceptance of the design. The engineer should realize that the design follows careful observations and consideration of environmental factors but that art meets science when a design develops in the mind's eye of the engineer.

REFERENCES

1. JM Dalla Valle. Exhaust Hoods. New York: Industrial Press, 1946.
2. New Jersey Department of Labor Bureau of Engineering and Safety Regulations for Exhaust Systems (no longer in force).
3. AD Brandt and RJ Steffy. Energy Losses at Suction Hoods. Trans. Am. Soc. Heating and Ventilating Engrs. 52:205, 1946.
4. WCL Hemeon. Plant and Process Ventilation. New York: Industrial Press, 1963.
5. P Drinker and T Hatch. Industrial Dust. New York: McGraw-Hill, 1954.
6. ACGIH. Industrial Ventilation 21st Edition. Cincinnati, OH: 1992. American Conference of Governmental Industrial Hygienists Inc., 1992.
7. EA Kleissler. Use of Computer Modeling for Depicting Air Flows in Tissue Mills. Tissue World 97:1997.
8. M Van Dyke. An Album of Fluid Motion. The Parabolic Press, 1982.
9. ES Ferguson. Engineering and the Mind's Eye. Cambridge, MA: MIT Press, 1992.
10. Unpublished data from Kleissler Company, Lakeland, FL.
11. MW First. Removal of Airborne Particles from Radioactive Aerosols. In Treatment of Gaseous Effluents at Nuclear Facilities. Harwood Academic Press.

9
Flexible Containment for Primary Manufacturing/Bulk Operations

Steven M. Lloyd and Ronald W. Wizimirski
ILC Dover, Inc., Frederica, Delaware

EDITOR'S NOTE

ILC Dover, Inc. utilizes soft goods technology to provide structural systems that are flexible and offer an alternate technology to standard rigid structures. The term *soft goods* refers to the innovative technology that allows the replacement of traditional metal and glass containment components. Dover specifically has developed the technology through such applications as the space suits provided to NASA and used by astronauts for planetary and orbital missions, and the landing impact bags used on the Mars Pathfinder Mission. ILC Dover has put to use this and similar high-technology applications to design and manufacture "flexible containment" solutions for the pharmaceutical industry.

This chapter focuses on flexible containment of compound transfer operations, using specific recent work that ILC Dover has completed in pharmaceutical primary manufacturing, or bulk operations. Also discussed is the time frame and talent involved for a successful engineering development effort. The author employs the DoverPAC™ System, one of the results from Dover's pharmaceutical development work, to exemplify the general containment concepts being discussed. In a broader sense, and as alluded to in Chapter 1, this chapter also exemplifies the strong evolutionary nature that the art of containment is still experiencing, and the existence of diverse technologies outside of pharmaceuticals being imported into the industry and linked together for novel applications.

I. INTRODUCTION

In the processing of pharmaceutical compounds, those compounds must typically be transferred to and from processing equipment, transfer systems, and storage containers. Typically, this is accomplished through a series of manipulative steps performed by plant operators. A transfer container, such as a Flexible Intermediate Bulk Container (FIBC), is placed on the discharge port of a batch processing vessel.

The processing and material transfer of hazardous compounds must be accomplished without exposing the operating personnel to the health hazards associated with the pharmaceuticals. Exposure to as little as parts-per-million–levels of the drugs (or less) can result in chronic health problems and even death. This requires the use of protective equipment, including full body suits and respirators. This equipment is cumbersome, hot, uncomfortable, requires time to don and doff, and must be washed and potentially incinerated after use.

Additionally, for two reasons, it is desirable that the transfer of such particulate materials be accomplished in a contained manner. First, it is often necessary to protect the compounds from environmental contamination. Second, the compounds being transferred are often in a very concentrated state, and the accumulated loss during a series of transfers can have a negative impact on the efficiency and economies of the process.

As an example of these challenges, consider the charging and discharging of intermediate bulk containers. The means for closing the top of filled FIBC conventionally has consisted of twisting the material closed around the top of the bag and tying it off with a wire or plastic closure. The optimum solution would be to heat seal the bags closed. However, the presence of both flammable gases and combustible dusts resulting from uncontained particulates creates the potential for explosion and has rendered the use of heat sealing impossible.

Powders, such as those employed in the pharmaceutical industry, are often used in conjunction with solvents and tend to agglomerate within a vessel. When the contents of a vessel are not properly or entirely discharged, operators in protective gear must facilitate the discharge. This, not only interrupts the process, but can pose a danger to the safety of the operators and contaminate the product.

Finally, certain conventional technologies employ stainless steel vessels equipped with hermetically sealed ports for the transfer and storage of pharmaceutical products. Such containers can, in many applications, be employed for transfer and storage. However, they are expensive to fabricate and must be thoroughly cleaned to remove all traces of the particulate prior to reuse in order to avoid cross-contamination. The cleaning and certification program is expensive and not entirely reliable.

As an alternative to stainless steel vessels, one conventional technology has employed a continuous tube of flexible material. While such a transfer container may be disposable, it must also satisfy the rigorous physical requirements imposed by the pharmaceutical industry. The containers must possess the required physical characteristics for the particulate transfer service, namely, antistatic properties, flexibility, and high strength. Conventional FIBCs, while exhibiting some of these characteristics, do not possess all of the properties required for use in a containment system as the one described in the following sections.

II. FLEXIBLE TRANSFER CONTAINMENT SYSTEM

The overall purpose of contained transfer systems is to provide an economical and reliable method for transferring compounds, facilitating the transfer without either exposure of the operating personnel and the environment to the compound, or environmental contamination of the compound. In the following example, this is accomplished by utilizing four approaches.

1. Employ a series of individual DoverPac™ liners, which are connected to a multiple o-ring canister assembly (see Figure 1). As one of these liners is filled, it can be removed without breaking the connection or seal. A continuous sleeve cartridge, which holds a series of transfer containers joined at the top and bottom in a long sleeve-like manner, can also be used. When the first container is filled, the area between the first and second container is heat-sealed closed, thus creating the top of the first container and bottom of the second container.

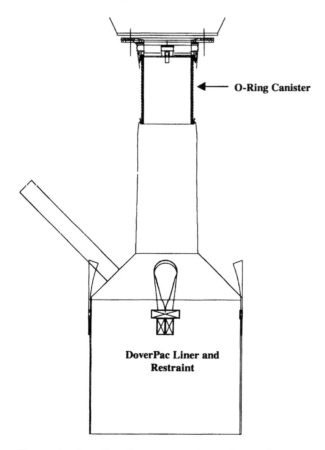

Figure 1 DoverPac liner connected to o-ring canister.

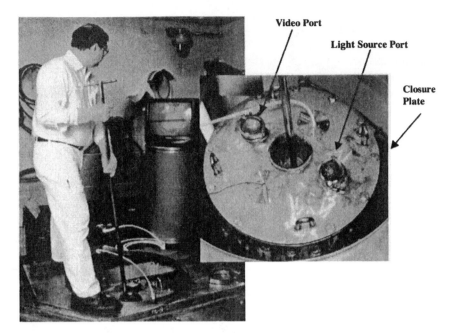

Figure 2   Rodding apparatus.

2.   Provide a flexible, thermoplastic, disposal transfer container material that is comprised of an olefinic base resin, an antistatic agent, and a filler. The material not only possesses the necessary combination of antistatic and mechanical properties for use in the contained transfer system, but can be fabricated in various configurations to meet the needs of a specific application.

3.   Provide a rodding apparatus that facilitates the discharge of compound that has agglomerated in a processing vessel (see Figure 2). The rodding system utilizes a closure plate allowing the rodding to be conducted in a contained manner. An integral long rod-and-blade assembly and two ports that allow the use of a video camera and fiber optics light source facilitate the rodding process.

4.   Develop a repeatable high-temperature heat-sealing capability that utilizes the necessary precautions to prevent the presence of arcs, sparks, and high-temperature exposued surfaces (see Figure 3).

The advantages associated with flexible containment are numerous. First, from an exposure standpoint, it provides for a shirtsleeve environment in which body suits and respirators are often unnecessary, enhancing both the operator's manual dexterity and overall comfort level. Second, from a containment standpoint, the system provides for improved economics resulting in reduced initial capital expenditure, along with enhanced recovery of airborne product and the processing of high-potency compounds in an uncontained processing facility.

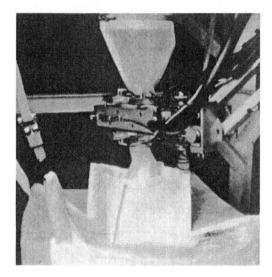

Figure 3 Heat sealer.

Finally, the flexible transfer container is a reliable, low-cost, and disposable alternative to conventional stainless-steel containers, while at the same time possessing the necessary combination of properties for use in the present system. For example, liners intended for high-potency parenteral drugs can be manufactured in a clean-room environment.

III. HARDWARE INTERFACES

One of the objectives of the flexible containment system is to provide a reliable and safe system for the transfer of compounds both to and from the processing vessels. Because there are a variety of types and sizes of processing vessels, a transition adapter is often bolted or clamped onto the vessel's charge or discharge flange and is used to standardize the opening of the vessel (see Figure 4). This stainless-steel adapter allows the use of standard-size hardware for the attaching mechanisms of these compounds and an easy retrofit of equipment currently in use.

In this example, a stainless steel disconnect flange is used to connect both the continuous sleeve cartridge and the multiple o-ring cartridge to the charge/discharge flanges of the processing vessel (see Figure 5). A series of quick-disconnect pins hold the disconnect flange in place. The flange also has a set of dual o-ring grooves to accommodate the change out of cartridges between production lots.

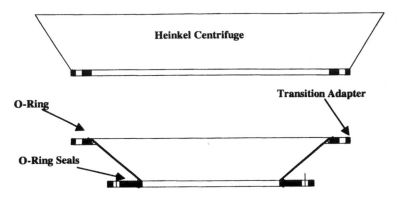

Figure 4 Transition adapter.

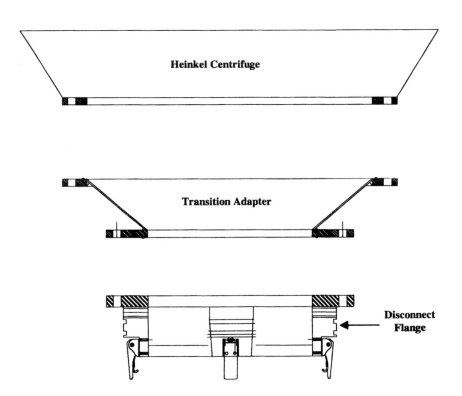

Figure 5 Stainless steel disconnect flange.

IV. DOVERPAC SYSTEMS

The DoverPac™ System exemplifies the equipment concepts and methodology being discussed. It employs two configurations of transfer containers: the multiple o-ring system and the continuous system. The multiple o-ring system can be used to either charge or discharge the processing vessel while the continuous system is used to discharge a vessel.

A. Multiple O-Ring System

The multiple o-ring system utilizes a multiple o-ring canister fabricated from an FDA-approved material such as polypropylene, and can be attached to the charging and/or discharge flange of the processing vessel (see Figure 6). The multiple o-ring canister is attached to the flange on the processing vessel by securing it to the disconnect flange with a gasket and a series of quick-disconnect pins.

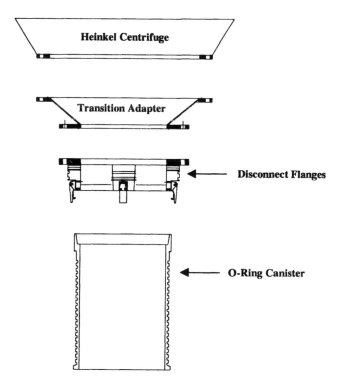

Figure 6   O-ring system.

Discharging the contents of the processing vessel comprises a series of steps. First, a liner, which provides the containment barrier, and its restraint, which supports the mechanical loads imparted to the liner from the compound's weight, are positioned under a processing vessel discharge flange. The liner's charging sleeve is attached to the first o-ring groove of the o-ring canister. The restraint is suspended from the bottom of the vessel by a suspension system that engages hooks suspended from the vessel. The canister connect sleeve is then connected to the vessel discharge flange adapter. The sleeve film stub ensures that the containment of residual particulate from the discharge flange is bagged out and secured in the sleeve. An inert gas, such as nitrogen, is introduced to fill the empty liner, thereby providing it with shape and facilitating its uniform indexing to the restraint.

The discharge valve on the processing equipment is opened, allowing product to flow into the liner. During this time, the sampling sleeve, if applicable, is placed into the particulate stream allowing it to fill (see Figure 7). The sleeve is then heat sealed or wire tied and cut, to separate the sample without contamina-

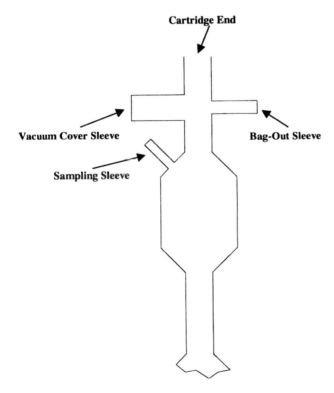

Figure 7 Sampling sleeve.

tion. After the first liner is filled, the top is heat-sealed closed and cut. The charging film stub remains on the bottom of the transfer canister, keeping the process equipment sealed from the environment.

A second liner is placed over the charging sleeve film stub of the first container and is secured to the second o-ring groove of the canister. The charging film stub from the first liner is removed from the multiple o-ring canister and contained in the bag-out sleeve (see Figure 7). The bag-out sleeve is heat sealed or wire tied, cut, and removed. The discharge process is then started for the second liner. This process is repeated until the entire contents of the processing vessel are dispensed, attaching the next liner to the next groove on the multiple o-ring canister. The process can be repeated for as many grooves as there are on the canister.

When the transfer operation is complete, the o-ring canister is disconnected from the processing vessel's flange adapter. The canister remains sealed because the bottom of the heat-sealed seam closes the top of the canister connect sleeve, and the top of the heat-sealed seam establishes the bottom of the canister connect film stub.

The steps for charging the processing vessel are similar to the discharging operation. The multiple o-ring canister with canister connect sleeve and the filled liner and restraint are positioned over a vessel-charging flange (see Figure 8).

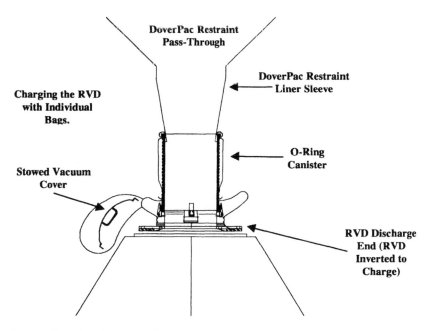

Figure 8 Charging a vessel.

The lower restraint is released to allow the liner's discharging sleeve to deploy. The discharging sleeve is secured to the first groove of the multiple o-ring canister. The canister connect sleeve is connected to the disconnect adapter with an o-ring. An expanding clamp on the top of the canister is actuated to prevent the accumulation of particulate on the top face of the canister. The discharge sleeve is untied and flow is established into the processing vessel. The rate of flow is controlled through a restraint cord around the discharge sleeve.

Once the liner has been emptied, the restraint is removed. The securing and expanding clamps are removed and the sleeve is closed by either heat seal, or twisting closed with tape or wire. The bottom of the heat-sealed seam establishes the top of the discharging sleeve film stub left on the top of the canister and the top of the closure seals the bottom of the empty liner.

A second filled liner and its restraint are positioned over the discharging sleeve stub left on top of the canister. The lower restraint is released to allow the second discharging sleeve to deploy and be secured with an o-ring to the

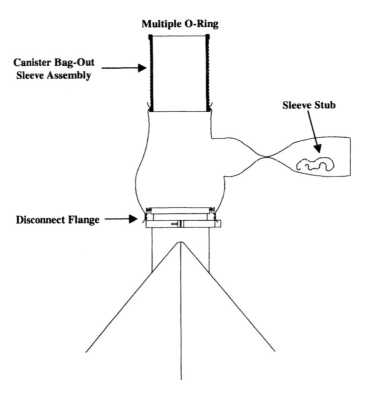

Figure 9   Removing discharge sleeve stub.

second groove on the multiple o-ring canister. The securing and expanding clamps are connected and the discharging sleeve film stub from the spent liner is removed from the top of the canister and placed in the bag-out sleeve (see Figure 9). The sleeve can then be tied off or heat-sealed closed. The process is repeated for as many liners that are needed to charge the vessel. The canister is removed using the same techniques for discharging a vessel. This method of discharging, connection, and disconnection has resulted in OEL levels of 0.2 to 0.4 $\mu g/m^3$ for a 12-hour time-weighted average (TWA).

B. Continuous System

The second configuration of the DoverPac™ system utilizes a continuous sleeve cartridge that, like the multiple o-ring canister, is fabricated from an FDA-accepted material such as polypropylene. The continuous sleeve cartridge assembly is comprised of a cartridge, a fabric shroud, and restraining straps. The fabric shroud envelops a sleeve-like series of individual liners that are joined top to bottom (see Figure 10). The sleeve of liners is compressed and packed around the circumference of the cartridge. Each liner comprises a sample sleeve and a bag-out sleeve. There is also a sleeve to accommodate a processing vessel's vacuum plate when required.

In a discharge operation, the continuous sleeve canister is connected to the discharge flange. If applicable, the vessel's vacuum plate is removed by

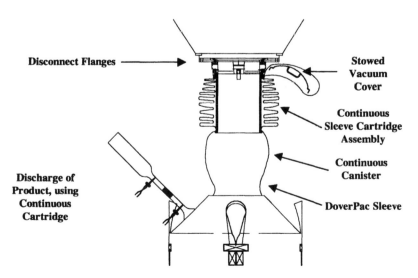

Figure 10 Continuous sleeve cartridge.

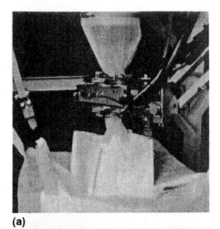

(a)

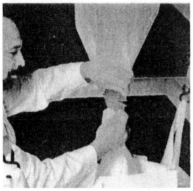

(b)

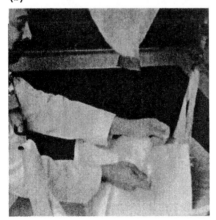

(c)

means of the vacuum plate sleeve and supported by a support sling. The cartridge connection sleeve is connected to the discharge flange on the vessel. The first liner is pulled down out of the cartridge and placed in the restraint. Nitrogen is used to fill the empty container and index the liner to the restraint. Then the discharge valve can be opened allowing particulate to flow into the first liner. As with the multiple o-ring system, a sampling sleeve can be utilized.

After the first liner is filled, the top of that liner's charging sleeve is heat-sealed closed and cut in the middle of the seal (see Figure 11). The top part of the seal becomes the bottom of a second liner that is now ready to be pulled out of the cartridge and placed into a restraint. The same steps are followed to fill the second and successive liners until the operation is complete. Throughout the filing process, the compound is contained, even when changing cartridges. Releases were less than 0.4 $\mu g/m^3$ for a 12-hour TWA.

V. RODDING SYSTEM FOR PROCESSING VESSELS

During the discharge operation, a rodding procedure is utilized to help break up compound that has bridged over the discharge valve or is stuck to the side of the processing vessel. The rodding system described in the following paragraphs is composed of a closure plate, retaining mechanism, tri-clover clamp, rodding poles, blade assembly, bellows assembly, video camera, and light source (see Figure 12).

The closure plate is installed on the top charging port of the processing vessel. A threaded post penetrates the plate and connects inside the vessel to the blade assembly (see Figure 13). The blade assembly is attached to the underside of the closure plate. The blade assembly comprises a blade and a connecting portion for engaging the post. A bellows assembly forms a barrier between the inside and outside of the vessel and allows the movement of the rod in the x, y and z axes (see Figure 13).

Figure 11 (Left) (a) Heat seal closure: A spent cartridge is disconnected from the discharge flange and the cartridge connect sleeve extends. (b) The cartridge connect sleeve is then heat sealed and cut. The bottom of the heat seal closes the top of the cartridge connect sleeve and the top of the heat seal establishes the bottom of the cartridge connect sleeve stub left on the bottom of the discharge flange. (c) A replacement cartridge has a connect sleeve that is connected to the discharge flange. This encapsulates the first cartridge connect sleeve film stub, allowing its removal through a bag-out sleeve. The bag-out sleeve is then heat sealed, cut, and removed.

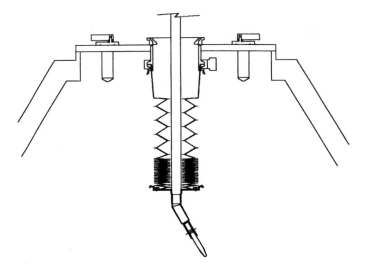

Figure 12 Rodding system.

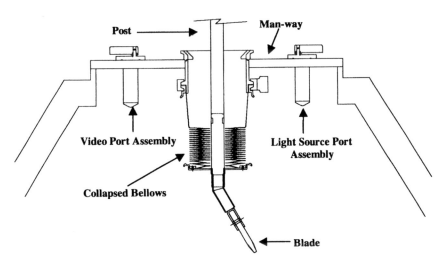

Figure 13 Rodding system bellows assembly.

The first section of a rodding pole is attached to the post. A means for securing the bellows and blade asemblies in a stowed position, such as a tri-clover clamp, is operated from the exterior of the vessel. This allows the blade to be deployed in a completely contained manner. When the clamp is released, the blade, post, and rodding poles can begin to descend into the vessel. Subsequent rodding poles can be attached to fully extend the blade assembly into the equipment.

A fiber-optics light source illuminates the inside of the vessel and a borescope video camera monitors the activity inside through a transparent glass view port. Figure 14 illustrates a view of a vessel port assembly that allows the insertion of a fiber-optic light source and borescope video camera. An operator moves the rodding pole up and down in combination with an off-vertical axis

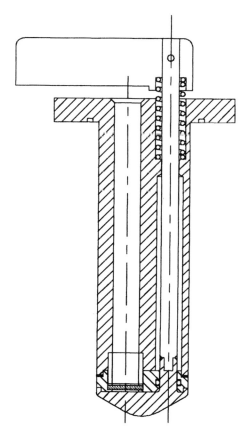

Figure 14 Vessel port assembly.

motion to scrape the inside surface of the vessel. This activity is monitored using an external video monitor that is connected to the video camera. When all contents of the vessel have been discharged, the rodding poles are retracted and removed from the vessel, and the blade assembly is secured to the top of the closure plate with a redundant retaining mechanism.

The rodding process can be repeated for subsequent lots of the same material without removal of the closure plate assembly. The closure plate is removed and cleaned between dissimilar lots of particulate. The bellows should be replaced during the cleaning operation.

VI. MATERIALS USED IN FLEXIBLE CONTAINMENT

A. Liner Materials

Liner materials, which are manufactured to meet the stringent standards required for use in the United States, Europe, and throughout the world, are fabricated from a flexible material that is comprised of an olefinic base resin film that is electrostatic dissipative. An antistatic agent and a filler are also added.

This type of film has excellent crack resistance and twice the strength and durability afforded by traditional polyethylenes. The antistatic agent results in very low electrostatic charge dissipation times, which is essential to the prevention of incendiary discharges as the liner is filled with particulate in hazardous locations. The liner material is thermoplastic, which allows it to be heat sealed to itself without the need for chemicals, primers, or adhesives. Additionally, the film is clear, which gives the operator the ability to view the contents of the container.

The liner film produces a thickness ranging from 4 to 6 mm; a surface resistivity (as determined by ASTM D-257-79, tested at 3,000 volts) of less than 1×10^{11} ohms per square; and a charge decay (as determined by British Standard BS7506) of less than 3.5 seconds.

Finally, the liner film can easily be fabricated in various configurations to meet the needs of a specific application.

B. Restraint Materials

The DoverPac™ restraint is constructed of a polypropylene using a plain weave construction. Electrically conductive yarns are woven into the fabric using a grid configuration of one conductive for every ninth yarn in the warp direction and every eighth yarn in the fill. The restraint is stitched together using pure x-static thread for conductivity and a three-cord polyester thread for structural integrity. Grounding loops are captured in every other corner and are made from tinned braided tube.

VII. HAZARDOUS LOCATION HEAT-SEAL MACHINE

A heat-seal machine for use in hazardous locations is depicted in Figure 15. There are five main components: the control enclosure cabinet, the user control box, the tower, the flexarm, and the heat-seal head.

The main control enclosure houses both the power and temperature controls and the logic circuits to perform the basic functions of heat sealing. It also houses two purge and pressurization systems to remove the presence, and prevent the ingress of, flammable gases and conbustible dusts.

The first purge and pressurization system is for the main control enclosure and is an automatic system that will purge any hazardous materials that may have entered the enclosure, and then pressurizes the enclosure to prevent further ingress of those materials.

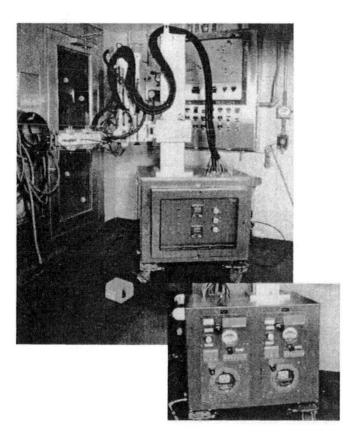

Figure 15 Heat seal machine.

The second purge and pressurization system is for the heat-sealing head and is designed to automatically purge and pressurize a small enclosure created by the heat-sealing head. This system prevents power from being applied or heat being generated until the purging and pressurization actions have been completed.

The heat-sealing head comprises several hinged bars that clamp together manually by the operator. The closure is sensed and automatically latched to prevent accidental or premature opening and exposure of the heating element to the surrounding atmosphere.

The heat-sealing head contains two temperature sensors, one on each side of the material being sealed. Operator controls are mounted on the heat-sealing head to allow the operator to operate clamping mechanisms on the head and begin the heat-sealing process. There are also control lights to indicate the status of the seal in progress or to warn of system faults.

The process is capable of detecting the temperature of the bottom side of the material opposite the heater. By knowing the temperature of the heater side and the temperature of the material side, and realizing that the temperature gradient through the material cross section is linear, it can be assured that the material interface being sealed is within a known temperature range. The system will allow the material to reach this level no matter how long or short the time is. Limits for the top and bottom temperature sensors ensure that the heater will not exceed the scorching limits of the material, and that the end of the process will be signaled when the set point is reached on the bottom.

VIII.   DEVELOPMENT APPROACH AND TESTING

A.   Development Approach

The success of these flexible containment systems can be directly attributed to the expertise of the product development team and creativity utilized in the development approach. The product development team assembled experts and the ultimate users of the system from both the pharmaceutical processing industry and the engineered flexible structures industry. Specifically, this team consisted of operators, containment engineers, process and design engineers, material development chemists, and industrial hygienists from the pharmaceutical industry.

The team developed a process flow chart illustrating the manufacturing operations and performance specifications for current flexible contained transfer systems. Utilizing continuous improvement and creativity tools, goals were established and different conceptual design approaches were formulated for each major processing operation. A combination of new technology and lessons learned enabled the team to identify and prototype several candidate designs. Evaluation testing and design trade-offs were conducted by the team, resulting in the state-of-the-art and user-friendly flexible containment system described in this chapter.

The development of the system was done in two phases with a total development time of eighteen months. The first phase was a six-month effort that included conceptual development, prototype evaluations, and demonstration tests on a full-scale mock-up at the development facility. The second phase of the project was a 12-month effort that included detailed design, fabrication of production hardware, and validation at the pharmaceutical manufacturing company.

B. Testing

To trade off competing design concepts and validate design approaches, testing was conducted at the flexible containment development facility on full-scale mock-up pharmaceutical processing equipment using a placebo pharmaceutical. The testing was conducted by the same product development team that generated the design. This approach gave direct, accurate, and timely feedback to the team, resulting in a design that met or exceeded all design goals. Testing was conducted to assess all functional aspects of the design, and contamination monitoring was conducted to determine the level of containment demonstrated by the equipment. Testing included discharging from a mocked-up centrifuge, charging a rotary vacuum dryer (RVD), discharging from the RVD, and rodding the RVD during the discharging. The remote video monitoring system was utilized during rodding trials. A fully functional heat-seal machine was used during all RVD charging and discharging evaluations. Airbone particulate monitoring and chemical-specific contamination monitoring were conducted during all evaluations to assess the level of containment being achieved during each operation. The system showed the capability of OEL of 0.4 $\mu g/m^3$ for a 12-hour TWA. The design was refined and upgraded based on test results and operator input. Direct operator feedback provided valuable information during development of the flexible containment system. The operators recommended design changes for the charging and discharging systems, the flexible containers, and the heat-seal machine. One specific example of this was the reiterative design of the heat-seal machine. The operators recommended that the heat-sealing head length be reduced from 26″ to 9″ to improve ergonomics, hinging the heat-seal head to make for easier loading of materials to be heat sealed, and reducing the overall size of the machine to improve operator visibility during loading and heat sealing.

A flexible containment system intended for actual production operations was fabricated for validation testing at the pharmaceutical production facility. Again, the same product development team participated in the validation tests. These tests were conducted using an actual high-potency pharmaceutical. Both wet cake and dry powder bulk compounds were tested. Relatively few problems were experienced during this commissioning phase due to the rigorous early development; testing; and the involvement of the user group through the entire development process. One example of a design problem and its solution had to do with the rodding system. Pharmaceutical material caked on the inside surface of the rodding plate cover and site glass covers preventing opening of the site glasses, and also prevented viewing and illumination through the site glasses. The site glass ports were reduced in diameter and redesigned to protrude three inches into the vessel. The new design was tested and verified, completely eliminating the problem. The system had releases of less than 0.2 $\mu g/m^3$ for a 12-hour TWA.

The primary reason for the success of this project was the development approach employed. When merging two very different technologies to address a highly critical and difficult challenge, involvement of the user group as a prime member of the development team is the key to success.

10

An Array of Containment Designs:
Following the Production Line in a Dry Products Secondary Manufacturing Operation

James P. Wood
Eli Lilly and Company, Indianapolis, Indiana

I. INTRODUCTION

In this chapter, you'll tour a production line to witness a range of containment approaches in current use. You'll see permutations of the basic plans that came about during one company's formative years of containment—process steps that are inherently self-contained within standard equipment, equipment that has been modified to become self-contained, non-contained processes that have been remediated in the field, and, finally, process steps that have not been contained at the source and rely on the surrounding facility or procedures to do the job.

No single approach can be guaranteed 100% effective for all sets of conditions. So, rather than a hard and fast set of recommendations, this chapter presents a set of working examples. For a compound with a fairly stringent Occupational Exposure Limit (OEL), the collection of containment methods and provisions described in this chapter have proven satisfactory for a unique combination of process, equipment train, and personal exposure conditions.

In some instances, you'll learn about the design process, the level either of design or of operational effort required for a designated level of containment. In other instances, a low-tech method could be retained with some modification, whereas in some cases a high-tech method was warranted. There are examples where containment could have been further enhanced and a discussion of how that could be achieved.

The design team generally favored self-contained, hard-piped equipment, and as little personal protective equipment (PPE) as possible. The team also realized that attaining these design goals, or even approximating them, would require in-house design effort, first to research the market and specify the appropriate equipment, then to modify certain equipment, install the process line, and commission the new facility.

II. THE PROCESS DESCRIPTION

A. The Product

The product being manufactured is a highly friable tablet, a formulation of a compound with an established OEL of 0.2 $\mu g/m^3$ over a 12-hour time-weighted average. The combination of product friability with a low OEL presents a double challenge in keeping airborne dust down and contained.

B. The Process

The process is a fairly typical granulation-to-tableting operation that is seen in many *secondary*, or *formulation*, pharmaceutical manufacturing plants, with these major process steps:

Dispensing of active ingredient
Transferring active ingredient into the equipment train, in this case the granulation vessel
Granulation
Drying
Milling
Blending
Tablet compression
Post-compression steps (metal checking, dedusting, inspection)
Bulk tablet loading

III. DISPENSING

A. Highlighting the Glove Bag

Let's start at the beginning of the process, the initial step of dispensing a 100% active ingredient from its bulk container. This is an operation that has greatly evolved from the early method and will serve as an example of the evolution of an individual operation and of a cost-effective-retrofit.

1. The Former Method

Dispensing was performed at a central dispensing facility, remote from the rest of the secondary manufacturing operation. A drum containing the high-potency active ingredient was towed to the designated dispensing suite, in this case, a full unidirectional air flow room. The drum was a fiber drum, nominally 18″ × 22″, with an internal liner containing the active ingredient. The drum lid was removed, the liner untied, and the compound hand scooped into jars approximately $4'' \times 7''$. The jars were then twice overpacked for transportation to the secondary manufacturing site for processing. Overpacking entails placing the jars in metal containers with a vermiculite-type packing material filling the space remaining within the metal containers. The containers are then taped shut and enclosed in plastic liners, corrugated liners, and corrugated boxes. The bulk of this packing is to comply with public transport regulations.

2. Containment Provisions

Containment was provided by unidirectional air flow. Operators wore two sets of PPE. The inner layer consisted of a tight-fit respirator and a full Tyvek coverall suit including connected gloves and booties. The outer layer was another full Tyvek coverall suit, with over-gloves and over-booties, and a Bullard breathing hood with powered breathing air supply. Note that the double layering of PPE extended to double breathing protection. The rationale for double layering was to protect the worker during degowning of the contaminated outer layer after dispensing. Gowning and degowning of the outer layer was done within the unidirectional flow room, with the room airflow remaining on. The contaminated outer layer was drummed within the room for later disposal.

Also notable was that this operation required two operators. One performed the operation, while the other observed the dispensing operation from an anteroom air lock separating the dispensing room from the general hallway. The second operator then helped clean the room and assist with degowning.

3. Containment Results

Both personal and area monitoring were performed. In the dispensing room, airborne concentrations were in the neighborhood of 100–300 $\mu g/m^3$. In the anteroom they ranged from 0.1–4 $\mu g/m^3$. Results are reported based in a 12-hour-time-weighted average format.

4. The Current Method

The dispensing step continues to be performed at a central facility.* The operation also continues to be performed in the unidirectional air flow room, this being the

---

* Whether dispensing is performed at a single centralized facility or is decentralized by plant or building reflects a company's philosophy and is driven by such factors as overall logistics of materials flow; distance to outlying plant sites; cost and availability of real estate; and other business and production considerations. The degree and reliability of containment is only one of many factors in that decision. Retention of a centralized dispensing facility at this particular site does not indicate failure to achieve the level of required containment but that other business factors figured into the design. Indeed, the level to which the containment effectiveness increases in an operation can actually enable a company to minimize or omit containment as a factor in selecting an optimal location.

space originally allocated for the step and because it is a secondary means of containment, although it is not needed under normal conditions.

Removal of compound from the drum is still by hand scooping, a decidedly low-tech approach for transfer (a high-tech method is not needed to achieve the desired containment). What has been added is a glove bag (see Figure 1). This has been designed especially for the operation and is a table-top model that allows an operator to work at a comfortable height. A sleeve attached to the glove bag docks onto the drum so that an operator can reach through the glove ports to perform the dispensing operation. PPE technically is not required for normal operations. A minimal level of PPE should be worn, however, in case of an upset (see the Cost Implications section).

The fiber drum containing active material is brought into the unidirectional air flow room. The drum is docked into the flexible docking sleeve in the glove bag and the sleeve end is taped, strapped, or otherwise secured around the top of the drum.

Reaching through the glove bag, the operator moves the drum lid, sets it aside inside the glove bag, and opens the liner. The operator then hand scoops compound into small jars for shipping to the secondary site.

5. Maintaining Containment in Between Dispensing Operations

Surplus compound is left in the drum to be saved until the next call for dispensing. This means that the drum must be disconnected and later reconnected to the glove bag without breaking containment. But once again, a low-tech approach was chosen. As long as compound is left in the drum, the docking sleeve end remains

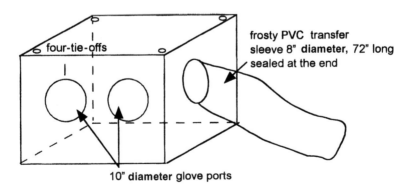

Small Dispensing Glove bag
Glove bag body: clear PVC
Glove sleeves: frosty PVC, 10" **diameter**, 20" long with
5 1/2" **diameter at glove end**

four-tie-offs

frosty PVC transfer
sleeve 8" **diameter**, 72" long
sealed at the end

10" diameter glove ports

Figure 1   In this example, a retrofit was effected by design and installation of a table-top glove bag within the laminar flow room.

secured around the drum. The sleeve is pinched a few inches above the drum, taped tightly, and cut through the center of the taped area to sever the drum from the glove bag (Fig. 2). Operators may use a dry or a wet cut, the wet cut being the more conservative choice, depending on the compound's relative hazard or dustiness index. More conservative still is an option to tape over the cut edges

glove bag sleeve

new sleeve over old

pull off rubber band, tape, or other attachment with the old sleeve stub, when new sleeve is attached over the old

Figure 2 Maintaining containment in between dispensing operations. (a) glove bag sleeve; (b) tape-and-sever disconnection method; (c) pull off rubber band, tape, or other attachment with the old sleeve stub, when new sleeve is attached over the old.

on both ends of the sleeve. A higher-tech method is to heat seal a section of the sleeve and slice through the seam. For this application, however, the simpler approach is entirely satisfactory.*

After the docking sleeve has been severed, the drum and sleeve stub assembly is reshelved in the warehouse until there is further demand for the compound. The assembly is then moved back to the dispensing room, and a new docking sleeve is drawn over the old sleeve stub and attached to the drum exterior. From inside the glove bag, the operator reaches the sleeve stub and pulls it off the drum, containing the interior drum emissions with the new glove bag. The old sleeve stub, now trash, can be passed out into a trash receptacle through another port in the glove bag.

Once enough dispensing operations have emptied the drum, the docking sleeve is left connected to the drum and the sleeve is cinched and cut from the glove bag. The sleeve stub is then collapsed and pushed down into the empty but still contaminated drum. Dependent on the hazard level of the compound, the operator can choose between the following disposal options:

1) Clamp a lid onto the drum top, sealing it over the sleeve stub and pinching the stub against the edge of the drum lip.
2) Load the empty drum and sleeve stub into a larger drum, possibly a 55 gallon size, and lid it for transport and disposal. This method is used for the production line described in this chapter.

The glove bag must also be removed after dispensing. The glove bag's HEPA vent filter makes depressurization possible so it can be collapsed. Then, depending on economic factors, the glove bag may be saved and reused, or it may be thrown away. Initially in this operation, the glove bag was collapsed and stored along with the partially filled drum of active compound in the warehouse until the next demand for that compound. As time went on, however, the glove bag was discarded after the dispensing operation. The cost of glove bags for this operation had decreased, partly because of increasing competition among suppliers and partly because suppliers became more efficient producers. Cost decreases were also driven by the customer's learning advances in this evolution, as design features of earlier glove bag designs were eliminated or modified.

6. Containment Results

To date, containment has been successful. Air monitoring that includes personal and area sampling indicates that airborne emissions are below the OEL. Readings ranged from upper values of one to two orders of magnitude below the OEL,

---

* This approach has also been used in certain applications within the nuclear industry, with good results.

down to nondetects, on a 12-hour time-weighted average.* Under normal operating conditions, neither respiratory nor garment PPE is required for the workers to operate under the OEL limits. In the event of an accidental glove bag tear, most likely a seam split, industrial hygiene recommendations are for one tight-fit respirator and a Tyvek upper jacket.

The production staff and the site industrial hygienist continue to log data from the facility to document a history on the reliability of the equipment and containment systems in place. As that happens, PPE may continue to be downgraded, as it already has to a significant extent.

B. Cost Implications

Areas of major cost savings for the switchover to glove bag use are as follows.

1. PPE

Without a Glove Bag. First operator: a full Tyvek coverall suit, boots, extra gloves, a HEPA tight-fit respirator, a second layer of Bullard breathing hood and powered air pump, Tyvek coverall suit, boots, additional sets of gloves. Second operator, same outer layer as the first operator.

With Glove Bag. PPE now downgrades to just one set of PPE: one tight-fit respirator, and one Tyvek upper jacket. Note that at this time, wearing even this PPE is a contingency judgment call of the health professionals responsible for the area. PPE technically is not required to operate under the OEL limits.

2. Personnel

The necessity for a second operator to audit the process and aid in the operation from time to time no longer exists. PPE requirements are significantly reduced due to greatly diminished airborne emissions. Procedures to assure operator protection and subsequent room decontamination have been simplified.

---

* Extensive air-monitoring studies have been performed in the dispensing area, both before and after source containment was implemented. The dispensing process, as described previously, is that of the actual compound being followed throughout this chapter. The dispensing data cited here is taken from air-monitoring results from a parallel study on dispensing operations of a different compound, this operation having a more extensive, robust monitoring database. The chief difference between the operations of the two compounds is in the quantities dispensed with each lot. This difference makes our conclusions more conservative, however. The dispensing operation, as described here, entails hand scooping compound into 4″ × 7″ jars, or 1–5 Kg quantities. The actual data presented here is of that same operation; however, the hand scooping is in the 15–25 Kg quantity range, significantly more compound being handled at one time.

3. Unidirectional Airflow Room No Longer Required As The
    Primary Means of Containment (Intangible Savings)

Originally, the anteroom was the only remaining backup to the general hallway
if that primary method (the room) failed to provide containment. The dispensing
room, by virtue of its unidirectional airflow, now serves as the secondary means
of containment. The room acts as backup and is not needed under normal circumstances. This allows the anteroom/air lock to be a tertiary barrier between any
emissions and the general hallway.

4. Cleaning

Without Glove Bag.   Cleaning of the unidirectional flow room occurred
between different lots of materials. Cleaning was to a visual level. Time required
amounted to 20–40 minutes per day, on average. All equipment and other surfaces, for example, scales, table, and other items were also cleaned. Additionally,
all cleaning materials and supplies (including mop heads, buckets, etc.) were
disposed of after each cleaning.

With Glove Bag.   The room no longer requires cleaning on a regular basis.
(It is still inspected between lots.) The scale platform is still cleaned, since it is
exposed to the same compounds, as it is inside the glove bag. There have been
a few occasions when a bag seam has begun to open, at which point the operators
revert to the earlier room cleaning methodology.

C. Glove Bag Design Pointers

1. In the infancy of glove bags in the pharmaceutical industry, off the
shelf glove bags were used that had been fabricated for such purposes as asbestos
abatement. This may have led to a stereotyped image of the device—a large
plastic cube with a couple of glove ports through which an operator can reach
in and presumably do just about anything. But maybe not too well, or too comfort-
ably. The company in focus has designed one-, two-, and three-, and up to seven-
handed glovebags, ranging in size from slightly bigger than a bread box to large
enough to enclose several sizable pieces of equipment. They have been square,
rectangular, triangular, octagonal, rounded, asymmetrical, and what-have-you.
The creativity that gives rise to such a wide variability follows the simple precept
that form follows function: the operator who is intimately familiar with the function goes through the motions to be performed within the glove bag as you observe and design. You'll see precisely how large the bag must be and how it will
be shaped to fit around the equipment. Then you'll get ideas for a supportive
structure. You may hang the bag on a metal or plastic framework or maybe it
will hang directly onto a piece of production equipment by elastic bands.

2.   Critical factors to think about are storage—where will tools be kept out of the way during glovebag use?—and trash removal—how will you get contaminated items safely out of the bag? How can the operator insert items into the bag without breaking containment should this become necessary? Extra sleeves may be offered for just-in-case use.

3.   Adequate filtration needs to be planned, not forgetting that air must be pushed out of the bag before disposal. Positioning of the filter should place it away from the operator's work area.

4.   Material of construction is a balance of sturdiness, clarity, and manageability. Does the fabric maddeningly cling to itself and the operator like a flimsy sandwich wrap? The product-contact acceptability may be an issue. Sealability is a critical element of durability. Can the fabric be sealed with dependable seam integrity? Depending on static arcing or explosivity concerns, conductivity of the material may be a consideration.

5.   Once your design is fairly well conceived, a quickly assembled mock-up from inexpensive materials at hand will expose major flaws in the concept. After this, your prototype should need only minor alterations for a workable production glove bag.

IV. ACTIVE ADDITION

A. Highlighting the Glove Box

Expertise in glove box design and construction has existed for many years and has found application in a variety of industries. It is beyond the intent of this chapter to offer complete design drawings and specifications for a glove box, but the reader may turn to any of many manufacturers highly skilled and experienced in designing and fabricating such units. What will be offered here is an example of how a glove box is used in the pharmaceutical industry, its integration with a process, and some tips germane to pharmaceutical manufacturing.

In this chapter's sample production process, the active ingredient is brought into a small room housing a two-chamber glove box with an attached side table, which is utilized for contained addition of active ingredient into the process (see Figure 3). The glove box, along with a small mixing tank located underneath, is the only equipment in this room, which is dedicated to this operational step. The active ingredient has come directly from the dispensing operation and retains its packing from that operation, here specifically, a glass jar. The jar lid has been screwed on and then taped around to prevent loosening during transit. The jar has been wrapped further in a cushioned bag, placed in a metal container, and, finally, boxed in a corrugated cardboard box. This last step is to comply with DOT regulations for transport over public roads.

Figure 3   A two-chamber glove box with an attached side table is used for containment of the active addition step.

Again, as with the dispensing step, an improved operation can be compared with the original to illustrate improvements in procedures. Here, the equipment has not been modified, but the location of one of the activities is changed.

1. The Former Method

The packaged compound was passed through the outer port door into the first chamber of the glove box. In the first chamber, the outer box and packing popcorn were removed and that material passed out the back of the unit through a trash port into a receiver bag. The primary containment device of the compound, specifically the glass jar's exterior, was now exposed to the interior of the glove box. It was passed through an interior port door into the second chamber, and the interior port door was resealed. The jar top was removed, contaminating the interior of the second chamber. Then the active ingredient was poured into a funnel/screen receiver, which fed into a vertical pipe through the bottom of the glove box and down into the next step on the floor below.

2. The Current Method

Based on experience gathered from the operation over time as well as air-monitoring data, the procedure was altered. Now active ingredient is unpacked outside of the glove box on the adjacent table. At that point, the glass jar is passed into the first chamber with the lid still in place. There it remains until needed, at which time it is passed into the second chamber for the addition step.

The operators and the management of the area have determined that the risk/benefit ratio for changing to this method is acceptable, given the added convenience and a history of receiving the container from dispensing with its exterior uncontaminated and unbroken; the track record of the operators not causing jar breaks outside the glove box; and the specific toxicology (hazard) of this compound. The added convenience stems from the operators' being able to unpack the material directly rather than through glove ports. In terms of time savings, the change amounts to 2–4 minutes per container. On this last point, even though the OEL is in the submicrogram range, the acute versus chronic toxicology profile is acceptable for shorter-term elevated exposures. This is another example of evolution in containment, in this case methodology, based on informed judgment derived from operating experience and specific hazard analysis.

Analyzing this operation, we see that the first chamber will remain uncontaminated during the process if two conditions are met.

1. The exterior of the jar is not contaminated when coming from dispensing—an example of how insufficient containment in one operation can create a negative impact in other areas, indeed, at an entirely different plant site in this case.
2. The inner port of the glove box seals tightly and uniformly, precluding leakage from the second chamber.

B. Cleaning/Decontamination

The glove box has a CIP nozzle within each chamber, and in both chambers the floor slopes to a drain in the center. Initially, a manual hose/wand was also located in each chamber, but these developed slow leaks because of unreliable shut-off valves, and the hoses have been removed. As experience proved the CIP nozzles adequate for all cleaning and decontamination requirements, removal of the hoses appeared to be best. Note the compound is not hydrophobic, but is nonsoluble in water so this is a middle-of-the-road cleaning challenge.

C. Glove Box Design Pointers

1. Construction details should be specified for ease of decontamination and cleaning. These include such considerations as coved interior corners and flush-mounted surfaces throughout.
2. Portal doors need to seal reliably. In this example, the doors are horizontal swinging with offset hinges. The closure mechanism and configuration needs to be such that equal pressure exists around all of the surrounding gasket

when the door swings shut. In analyzing the points of door-to-gasket contact, the inner door swing tends to contact the gasket before the outer door portion, and over time this can cause uneven wear on the gasket, a detail that can undo the rest of a well thought-out design.

3.   Criteria for the ventilation and air filtration system must be thought through with each glove box installation.

In this examle, the air circuit is a one-time pass-through route. Make-up air is drawn into the glove box from the room through a back-mounted HEPA filter assembly. Note here that filtration of inlet air to the glove box is not required for quality reasons; the ambient room air has already undergone the degree of filtration required for the compound. The glove box HEPA filter provides a source of make-up air to the glove box and at the same time protects the room environment from contamination internal to the glove box.

The HEPA filters need to be changed periodically. How can this be done in a contained fashion? During the design phase, the frequency of filter changes could not be accurately estimated, because there was little experience in applying this containment approach to the specific compound and process. The HEPA filter would need to be changed during actual processing operations while the glove box was contaminated. With this in mind, the filter configuration selected was a push-through arrangement. The filters are round and fit inside a metal sleeve penetrating the glove box wall. When one filter is to be changed, a clean filter is pushed from the outside into the metal sleeve. Further pushing into the sleeve forces the dirty filter out into the glove box for disposal into the trash port. Not all processes require this degree of contained filtration change capability. Indeed, in this installation, later experience indicated that the filters load slowly enough that the change can be planned for a time when the chamber has been decontaminated. However, if that information or experience is not available during the design stage, be aware that contained filtration change provisions are available in the marketplace, be they push-through configurations as previously mentioned, bag-in "safe change" assemblies, or other designs.

Another consideration is moisture. Water and HEPA filters do not mix. Hot water from the CIP system can create a significant amount of steam that, when contained within the glove box, can find its way to the HEPA and ultimately blind it. In this installation, the glove box ventilation system is set to different operating parameters during the CIP cycle to minimize steam contact. Additionally, there is an integral shroud arrangement over the HEPA filter face on the glove box chamber side.

4.   A visit to the fabrication shop to see the progress of the glove box construction is advisable, especially when the glove box will be a major part of the overall installation. In this example, an intermediate visit to the shop showed that the glove box floor didn't slope to its drain. Because of the visit, the problem

was corrected prior to the entire assembly being put together and shipped to the owner, saving significant time, expense, and avoiding consternation.

5. Don't just deposit the operators next to the finished mock-up walk off. They will be, quite understandably, unable to offer much in the way of useful input. By the time a mock-up is completed, the operators should have been brought into the conceptualization process, offering their input, and ultimately buying into a rationale of why a glove box is being used versus the standard of wearing PPE or "the way we always did it."

6. Also during the mock-up stage, remind everyone that the mock-up should be seen as totally changeable, a "cartoon" that can potentially be stretched, bent, or reshaped as desired. Is the glove port too high? Lower it. Is there an obstruction in the field of vision? It can be moved very easily at this stage of the game. The operators should physically perform the simulated operation to identify ergonomic conflicts and aid the design team in working them out. Does this mean more than one mock-up might be constructed? Possibly. But it's much less expensive to cut and patch plywood and plastic than stainless steel and glass. One of the defects identified in this particular example was the lighting within the glove box, which was causing a reflective glare in the operators' eyes. Optional solutions included a more diffuse lighting source, a slight modification of the angle of metal surface, or a different finish on the reflective metal surface.

1. Containment Results

Personal and area air sampling have indicated that the environment in the area is significantly below the OEL for the compound. Operators are "in shirt sleeves" during both operating and cleaning processes, while handling the active ingredient at 100% strength.

V. GRANULATION/DRYING

A. Highlighting Self-Contained Processing Equipment

Combining multiple processes within the same vessel or enclosure is generally a step in the right direction, reducing transfer points through the process train. If the process were all hard piped and remained 100% closed, combining steps would theoretically be irrelevant to containment. Yet theory has a habit of not translating to fact in the field, so physically reducing the number of potential emission sources, through reduction in vessel transfer points, is generally a plus for containment. Planning needs to take into account other process considerations, but all other things being equal it's an idea worth considering.

Figure 4 In this operation, granulation and drying occur in the same vessel. The piece of equipment utilized is a combination granulator/microwave vacuum dryer.

Continuing with our production line, the active ingredient is next piped into the granulator/drying vessel (Figure 4) from the active addition glove box located in a separate room directly overhead. Excipients and granulating solution are added in this step as well. All ingredients are piped into the granulation vessel with no disconnections. The vessel does have an access port in the top, and much attention has been given to the effectiveness of its gasketing material and the uniformity of seating against that gasket when the access port closes. Decontamination is achieved via clean-in-place systems after processing is completed.

Early in the project stages, the process development staff was concerned that damp powder cake might stick to the underside of the lid and the sides of the bowl such that it would be necessary to open the lid periodically and manually scrape down the powder. Obviously, this would be a significant containment problem. The development and pharmaceutical engineering staff tackled this issue directly from a process and equipment standpoint. The vessel was jacketed, so side wall temperatures were modulated to approximate the temperature of the compound inside. Also, the internal blade was reconfigured for tighter tolerances and a different imparted-energy profile. These actions greatly reduced the residual compound within the vesssel. In this way, the process was directly improved while eliminating the potential containment problem.

The process equipment is self-contained during both manufacturing and cleaning. Both the manufacturing and the cleaning steps described here are performed with no PPE requirements for the operators.

VI. MILLING

A. Highlighting Self-Contained Equipment, With Contingency Design Concept for Secondary Containment if Needed

The milling operation in this example could rightfully be included in the granulation and drying section. In a fairly standard configuration, the mill is secured onto the side of the granulation vessel and is for all intents a part of that single-vessel configuration. The mill, however, can present unique containment challenges apart from the granulation or drying vessel feeding it.

1. The Process Flow

The process flow through this segment is straightforward. Once the compound has reached its final state in the vessel upstream—in this case, once drying has been achieved—the compound is automatically and continuously fed into the mill inlet, through the mill, and onto the next process step.

2. Containment Challenges and Concerns

The piece of equipment is an offset screen mill (see Figure 5). Placebo tests that were run on the equipment prior to installation to assess its relative containment effectiveness were very encouraging. But the design team recognized that the tests by their nature were valid for the short term only; longer-term scenarious that couldn't be tested for remained causes of concern.

> Residual compound could build up in the interior of the drive mechanism. The mill is belt driven, and its shaft penetrates the otherwise closed housing of the mill. Sealing between and within the shaft bearings through the housing is a potential leakage point. If leakage were to occur, over time residual would migrate through the housing penetration and be carried along the belt, contaminating the interior of the belt guard and motor, which are not contained pieces of hardware. Compound residuals would then become open to the room environment.
> The mill housing's tight closures and gasketing might prove to be short lived due to premature wear, uneven seating pressures, or other factors.

Figure 5   The mill and blender.

If so, the mill would leak under normal production conditions, which is an unacceptable condition. A contingency design was developed that was essentially a small glove box/glove bag hybrid. The rigid, or glove box, portion of this enclosure was designed to be located across the back, bottom, and either one or two sides of the mill. The remaining sides would be flexible, with glove sleeves allowing operators to reach in for adjustments. The materials and methods of construction for the rigid portion would not be to the same standard as the glove box in the earlier active addition step. Here, the rigid portion would serve mainly to support and stabilize the flexible portion.

3.   Cleaning

The mill is cleaned via CIP, during the same cycle that cleans the granulator/dryer feeding it.

4.   Containment Results

The process equipment is self-contained during both manufacturing and cleaning. Both the manufacturing and the cleaning steps described here are performed with no PPE requirements for the operators.

In some other installations, a mill is sometimes undersized for the product flow being fed through it. In these instances the mill will choke, requiring it to be opened manually to dislodge the potent compound. This can cause emissions that can make the sum of all other emissions throughout a well-designed system pale in comparison. Sometimes the mill can choke—not because of overload, but primarily due to the nature of the formulation being fed into it. There are two different approaches to this that bear mentioning. The first, modify the formulation to reduce any bridging tendencies, is process oriented. (Undertaking this effort is not trivial, but it does get to the heart of the actual problem and solves it long term. Tech Service and possibly Product Development areas will be of help here.) The other approach is remedial, and that is to add an extra push to the compound as it travels through the mill, keeping it in a more fluidized state. Two or three well-placed compressed-air outlets, for example, aimed at strategic targets within the mill housing interior, can prove useful. This approach or its variants, while not ideal, will still be preferable from a containment perspective to opening up the mill.

VII. BLENDING

A. Highlighting Self-Contained Equipment With Remedial Modifications and Local Exhaust

During the design phase of the project, the choice of blending technologies was now narrowed to two: a high-shear, or "ribbon," blender or a tumble blender. Much of the selection process centered around containment. In the tumble blenders favor was that the bin should be leak tight, but the transfer line that feeds product into the bin would need to be disconnected, and this was a major emissions concern. The high-shear blender required no disconnections because it remains stationary. On the other hand, the high-shear blender had a higher probability of leakage than the tumble bin because of the tolerances and long-term rougher handling usually seen with the access ports of this type equipment.

The project team decided on the hard-piped, no-disconnect methodology of the high-shear blender, believing the access ports could be modified in the field to a leak-tight condition. At the time of the project's design and equipment specification, the team was unaware of a method of contained disconnection that would have assured containment of the tumble bin. No significant leakage data was available at that time to help compare the two pieces of equipment so the team's best engineering judgment relied on past experiences and new product literature.

1. The Process Flow

As the product is discharged from the mill, it falls through its transport line into the blender. Other excipients are added at this stage as well. The added quantities are small, generally several kilograms, and are fed easily into the blender manually. After sufficient blending has occurred, as determined by noninvasive methods that allow access ports to remain sealed, the blender's bottom discharge valve is opened and the mixture is gravity fed down to the next step.

2. Containment Results

Air monitoring indicates that the operation regularly remains below the OEL threshold for the product.

3. Improvements

The step where excipients are manually added to the blender is a potential emissions source. It was judged that a well-engineered local exhaust would be a satisfactory remediation. At this stage in the process, excipients have diluted the active ingredient to 1% or less of its original concentration. That alone does not sufficiently reduce the potential of exposure for compounds of this level of potency. But the excipients were poured into a funnel/screen assembly configured so that once an initial amount of excipient is charged into it, it completely blocks the pathway from the interior to the exterior of the blender until the funnel clears. Added to that, the blender is under mildly negative pressurization for processing reasons. Finally, the local exhaust was designed to capture errant dust rising to the lip of the funnel, even though such dust would be nonpotent excipient being poured in rather than the active ingredient "trapped" inside the vessel. There being several modes of containment embodied in the design, containment at this location is satisfactory for the operation.

The production area is currently contemplating replacing the high-shear blender with a tumble blender, for process reasons irrelevant to containment. Designs and knowledge have evolved. Since the original design, contained methods of connection and disconnection have been developed and tested, and disconnection requirements need no longer play a major role in choosing options for this operation.

VIII.   TABLET COMPRESSION

A.   Highlighting Process Self-Containment Within Equipment After Modifications

1.   The Process Flow

Blended powder is gravity-fed through the floor of the blending room, entering through a metal sleeve penetrating the ceiling of the compression room below.

A flexible, double-ply polyethylene sleeve is connected to the metal sleeve immediately below the compression room ceiling. The lower terminus of the sleeve is connected to the inlet of the tablet press, where the compound enters the press and is formed into tablets.

2. Containment Challenge

Tablet compression traditionally hasn't been known either for its tidiness or its containment. Multi-ton hammer compression of 1,000–3,000 tablets every minute tends to generate errant powder. This compound is extremely friable to begin with, so this operation could generate dust quickly. Both manufacturing and cleaning steps required attention here because dust could be emitted during either.

3. Containment Features

A tablet press manufacturer whose equipment fit the project's needs was identified. The project team worked with the manufacturer to review the standard internal containment features and then to design other containment features into the machine (see Figure 6).

Standard Features.

Air-pressurized neoprene channel gasket around each access door, for positive sealing while closed.

Internal exhaust slots, engineered for efficient pickup at major dust generation points within the compression chamber.

Interior compression chamber kept at negative pressurization.

Modified Features. Glove ports located and installed in the access doors of the compression chamber, with closures and safety switches for automatic carousel lockout in the event any of the closures were opened. The team realized that the time of highest emissions would be when the chamber side doors were open for operators to make minor adjustments during a process run or for minor or major cleaning.* Glove ports permitted these operations to be performed with the doors closed.

---

* *Minor cleaning* refers to manually vacuuming visible powder and stray tablets from inside the compression chamber; removal and wiping of punches; and wiping and using a methanol swab and brush in the cavities left in the die table. This is done between runs of the same product.

Major cleaning refers to all minor cleaning activities plus full decontamination and cleaning of the carousel and remaining internal surfaces. This is generally performed between lots and after final production runs. Major cleaning also entails wet mopping the room's floor and walls, as well as moving mobile equipment such as the deduster and metal checker to an adjacent cleaning area. Major cleaning for this room and tablet press typically requires up to an hour, excluding the time to set equipment back up.

Figure 6 The tablet press was a major focus of concern for the design team.

Anticipated But Not Done.   Positively pressurize the upper and lower electronic and mechanical chambers of the tablet press to keep powder from infiltrating to these areas.

Unaddressed Emission.   During major cleaning, the carousel must still be exposed to the open room environment, and this is an obvious source of room contamination. After carousel cleaning, periodic room decontamination occurs. The team considered spraying the inside of the compression chamber with a CIP-type nozzle and other similar concepts, but were concerned over corrosion of the carousel, constructed of mild steel for better long-term wear during die-punch impact.

Tablet press manufacturers will now discuss the differing levels of clean-in-place capability available on their equipment. Because major cleaning of the press remains the largest factor in external emissions, focus on internal cleaning capabilities for future installations is warranted.

4. Changes Made in Procedure and Hardware Since Startup

Subsequent operational experience with this piece of equipment has demonstrated that minor cleaning and in-run adjustments can be performed with the access doors remaining shut as designed. These adjustments are made to tooling, instrument settings, or other hardware during the run cycle of a product lot. But with the modifications to the equipment as installed, it is easier and less time consuming for the operators to simply don PPE and open the access doors for minor cleaning. The PPE consists of a tight-fit filter mask only, not the more substantial powered breathing apparatus or air lines and full body suit. The latter PPE is not needed because the access doors are open only a short time and lower levels of emissions are generated. The operators, based on their experience, believe that the following changes in the machine's containment provisions would increase the practicality of keeping the access doors closed during cleaning.

1) Install a separate dedicated port through one of the access doors for removal of punches and small tooling.
2) Select gloves for maximum tactile sensitivity.
3) Provide a "better reach," meaning in this case greater latitude in horizontal movement for arms when reaching.
4) In some cases, reposition glove port openings (through the access doors) from those locations initially selected.

Implementing some of these modifications may not be practical. Specifically, the punch removal process entails repetitive back-and-forth motions as the operator reaches into the chamber to remove punches, then moves his arm out to push a control and index the carousel around, then reaches back into the chamber to remove more punches, and so forth. Thus the operator would be constantly pushing into and pulling out of the glove, significantly increasing the task time. A modification to consider is to install a remote switch inside the compression chamber, accessible without putting on and taking off of the glove. For safety reasons, there would probably need to be two such switches, each of which needs to be pressed simultaneously prior to the carousel indexing, assuring that the operator's hands would be clear of the revolving machinery.

5. A Facility Note

The room that houses the tablet press acts as secondary containment for the operation. Equipment in the room includes not only the tablet press, but also the metal checker, deduster, and sometimes a tablet inspection machine.

6. Containment Effectiveness

The operators are not required to wear PPE for this operation, but this is some-what misleading because the process is largely automatic. The tablet press is housed in a process room, and is automatic to the degree that allows operators to be absent for a majority of the run time. The main functions for which operators must be present are taking tablet samples, testing those samples, and cleaning the equipment afterward. With the exception of cleaning, operators' potential exposure levels are below the OEL during these times as well without use of PPE. This points to a two-pronged approach to minimize PPE requirements for the operators. On the one hand, if the process, hardware, software, and layout can be designed such that operators do not need to be in the presence of potential emissions to begin with, the requirement for PPE during processing goes away, regardless of how contained the operation. On the other hand, even when the operators' presence has been minimized, the designer should strive to contain at the source as much as possible. In this example the machine is still significantly contained during operation, as described previously, and somewhat contained during cleaning. This precludes a buildup in residual active ingredient over time throughout the room. This last point cannot be overstressed. If a process is allowed to "leak as much as it will" following the mind-set that it will be cleaned up afterward each time, experience shows that long-term background residual levels of the compound will rise to the point where PPE may be required even upon entering the area with no production being performed.

IX. POST-COMPRESSION PROCESSING

A. Highlighting Simple Localized Barriers and Minor Negative Pressurization as the Means of Primary Containment

After tablet compression, the tablets are routed through dedusting and metal checking equipment (see Figures 7 and 8).

1. Containment Features

The deduster and metal checker are good standard pieces of equipment and have not been significantly customized or modified. From a containment standpoint, some perspective in required here. At this point in the process, the compound is no longer a loose, bulk formulation, but a compressed tablet. Some errant powder may be generated, but at magnitudes less than in earlier formulation steps. However, this particular compound is purposely formulated to be very friable. Dusting from this point in the process still cannot be assumed to be negligible.

As in some earlier process steps, the low-tech approach was the one of

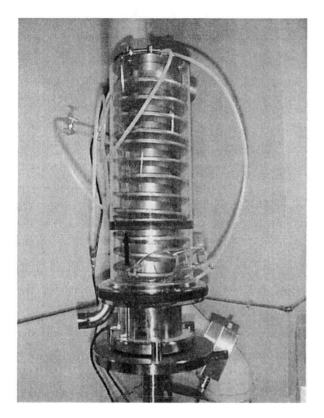

Figure 7   Tablets pass through the deduster after tablet compression.

choice. First, the transfer route that the tablets take is a connected, unbroken path from the tablet press discharge, on through the deduster and the metal checker. Traditionally, this pathway is some variation of an open-trough arrangement or similar hardware. Transparent covers were fabricated to lay over the troughs throughout the entire route beginning at the tablet press discharge. This causes two things to occur. The obvious one is that the direct line-of-sight pathway from the compound to the operators' breathing zone is shut off. Secondarily and a bit more subtly, the tablet machine, deduster, and metal checker now function as a single unit in terms of their barometrics. Once a constant negative pressurization has been established within the compression chamber of the tablet press, makeup air for that negative pressurization will come from the room, entering via the last piece of post-tableting equipment that has been overshielded by a cover. In effect, the entire tablet transfer route comes under the negative pressure pulled from the tablet press.

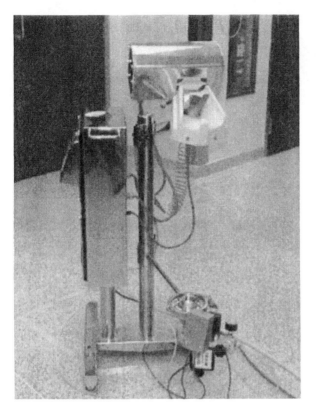

Figure 8   After dedusting, the tablets go through a metal checker.

X. BULK TABLET LOADING

Here, tablets that are discharged from the last post-tableting step are fed into a Venturi-type pneumatic transfer tube and conveyed to a location directly outside the contained processing area. Tablets are then gravity fed into pails of between 10,000 and 50,000 tablets per pail.

Certainly, there are methods for enhancing containment at this step. The point is that it isn't needed here. The terminus of the pneumatic tube that the discharged tablets fall from is also under negative pressurization from the Venturi system. With judicious positioning of the tube terminus with respect to the upper lip of the receiving pail, the incoming air sucked into the tube acts as an effective local-exhaust point, capturing any stray powder arising from tablet transport. In fact, prior to construction, model testing was performed to determine optimal locations for the tube position relative to the pail and operating parameters required for the Venturi system to transport these particular tablets without breaking or powdering.

The advantage of filling the pails in the general area is that external contamination from a potentially contaminated environment does not occur, as it would had the pails been filled within the containment zone. Saved time, otherwise required to decontaminate the outside surfaces of the pails prior to transport into the general area, can be used for other activities.

A. Containment Effectiveness

Subsequent monitoring confirms that this operation runs below the OEL threshold.

11
Containment System Selection

David Palister
Extract Technologies, Somerset, New Jersey

EDITOR'S NOTE

In this chapter, we explore how one of the industry's major players, Extract Technology, an equipment supplier, has significantly standardized a detailed assessment program for the purpose of project-to-project consistency—a laudable but often elusive goal. Extract's assessment program is delineated here not necessarily as a prescriptive standard but as a useful guidance tool for selection of containment approaches appropriate to particular operations. Other organizations have differing selection guides, based on varying degrees and types of data. The commonality in all these guides lies in an analytical approach, recognizing that degrees of containment legitimately exist and need to be matched to the hazard at hand.

I. INTRODUCTION

Many, if not most, pharmaceutical and chemical manufacturers already employ some degree of containment technology to protect personnel from exposure to harmful substances. Beyond the ethical and legal obligations to control and contain these substances, regulatory compliance and personnel insurance concerns demand scrupulous attention to containment. Wise choices among containment equipment, methods, and training should accomplish safe handling of hazardous materials under specific task conditions and should meet current industry standards, while anticipating more stringent requirements that may be called for in the future.

Standard containment practices across chemical and pharmaceutical industries, however, do not exist, and guidance is often needed to help these companies make wise choices from the containment equipment currently marketed. Equipment ranges from off-the-shelf local extraction arms to the complex glove boxes that maintain total product isolation. What are the criteria for making selections along such a range of equipment? A local exhaust arm, for example, may be low cost, flexible, adaptable, and mobile, and may do well to capture emissions during powder handling. But would an enclosed system offer a better solution for the long term? This chapter offers a control strategy selection guide based on the unique factors of a production operation.

For the purpose of our selection guide, a control strategy is defined as a specific device or engineering control for containment of dust or vapor during production activities. This includes good operational techniques as well as containment devices. Control strategies in the selection guide we present are placed along a continuum of hazard protection beneath a pyramid of factors representing production conditions that point to the appropriate control strategy. The factors that figure into the control strategy selection guide are

1. Operator interface and duration
2. Scale, or quantities being processed, and
3. Exposure potential band (properties of the compound and classification according to an exposure limit)

II. OPERATOR INTERFACE AND DURATION

An exposure profile that is a typical pattern of peaks of exposure to dust or vapors that occur at specific times during an operating cycle can be determined for each operation—for example, an instance of vessel charging or disposal of an empty sack. The operator-to-material interface cannot be precisely quantified, but certain commonsense evaluations can be made: For operations performed on a monthly schedule, consider the lower end of the containment strategy selections. For continuous operations with intensive operator interface in large-scale production, you should move toward equipment at the upper level of selection.

III. COMBINING PROPERTIES OF THE COMPOUND WITH SCALE TO FIND AN EXPOSURE POTENTIAL BAND

After considering operator interface and duration, you'll look at the nature of the compound, whether solid or liquid, and the amount being processed.

A. Solids

Dustiness of a solid is a factor in evaluating the degree of exposure an operator may face. The table below categorizes solids as high, medium, or low according to the contribution of airborne dust to exposure potential:

High	Fine, light powders. When used, dust clouds are seen to form and remain airborne for several minutes.
Medium	Crystalline, granular solids. When used, dust is seen, but settles out quickly. Dust is seen on surfaces after use. Example: soap powder.
Low	Pelletlike, nonfriable solids. Little evidence of dust during use. Example: PVC pellets.

The dustiness factor combines with scale of use, or the amount of a compound being processed, to arrive at an exposure potential band, a ranking for the hazard of a particular compound in a particular operation. The exposure bands for solids, or EPS for exposure potential of solids, are given a numerical ranking in Table 1.

Table 1 Scale and Dustiness Exposure Potential Bands

EPS ranking	Scale	Dustiness
EPS1	Gram quantities	Medium/low
EPS2	Gram quantities	High
	Kilogram/ton	Low
EPS3	Kilogram quantities	Medium/high
EPS4	Ton quantities	Medium/high

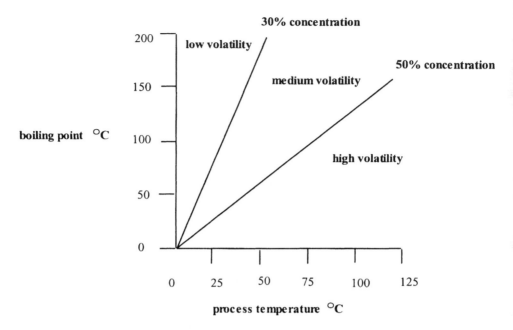

Figure 1 Exposure potential for liquids.

B. Liquids

As with solids, the exposure potential for liquids is affected by the scale of operation, but the dustiness factor is replaced by a category of volatility (Fig. 1). Thus,
an EPL, or exposure potential of liquids, similarly to the EPS, can be ranked as
in Table 2, again by a combination of scale quantities and a hazard level associ-

Table 2 Volatility and Scale Exposure Potential Bands

EPS ranking	Scale	Vapor pressure
EPL1	mL quantities	Low
EPL2	mL quantities	Medium/high
	m³/liter quantities	Low
EPL3	m³ quantities	Medium
	Liter quantities	Medium/high
EPL4	m³ quantities	High

ated with the compound, in this case the vapor pressure (VP), to arrive at an exposure potential band for an operation.

IV. MATERIAL HAZARD

Just as you determine exposure potential bands according to dustiness or volatility in combination with scale, you will determine the compound's health risk, this time checking with the Material Safety Data Sheet (MSDS) for the occupational exposure limit (OEL), possibly using an R-Phrase (the R-Phrase is a European list of hazard attributes that one can identify and assign to a compound when the exposure limit has not been determined) where the exposure limit cannot be ascertained. Assignment of an occupational exposure limit is typically done by the plant safety department or industrial hygienists, sometimes in conjunction with the research or toxicology group associated with the compound in question. The Institution of Chemical Engineers of the UK endorses use of an occupational exposure band for categorizing material hazard that we use in our selection guide. Following is the framework for assessment developed by the UK Health and Safety Executive.

Exposure limit range solids	Exposure limit range liquids	Hazard band
10,000–1,000 $\mu g/m^3$ dust	500–50 ppm vapor	A
1,000–100 $\mu g/m^3$ dust	50–5 ppm vapor	B
100–10 $\mu g/m^3$ dust	5–0.5 ppm vapor	C
10–1.0 $\mu g/m^3$ dust	.05–0.05 ppm vapor	D
1.0–0.01 $\mu g/m^3$ dust	0.05–0.005 ppm vapor	E
Below 0.01 dust	Below 0.005 ppm vapor	F

Where multiple products are to be handled, one would typically design a system applicable to the worst-case material. But it must be emphasized that designing a system based on an OEL alone is incorrect. Operator exposure may be greater from lower hazard dust in high quantities compared to a pharmaceutical active handled in grams over a short duration. Material type, operation, task time, and throughput all significantly affect exposure.

The factors affecting your choice are collected in Table 3 as an information gathering checklist, listing sources of input that have proven in the past to be excellent.

Table 3 Information Gathering Checklist

Elements	Chemical supplier	Occupational hygienist	Health and safety development	Operator group	Maintenance group	Quality assurance validation
Hazard Band OEB	✓	✓	✓			
Scale of operation				✓		
Exposure potential		✓	✓	✓	✓	
Frequency & task duration				✓	✓	
Operability of device		✓	✓	✓	✓	✓
Cost of device				✓	✓	

Table 4 Matrices for Finding a Control Strategy Level

Hazard band	Exposure potential solids			
	EPS4	EPS3	EPS2	EPS1
A	Control strategy 2	Control strategy 1	Control strategy 1	Control strategy 1
B	Control strategy 3	Control strategy 2	Control strategy 2	Control strategy 1
C	Control strategy 3	Control strategy 3	Control strategy 3	Control strategy 2
D	Control strategy 3	Control strategy 3	Control strategy 3	Control strategy 3
E	Control strategy 4	Control strategy 4	Control strategy 4	Control strategy 4
F	Control strategy 5	Control strategy 5	Control strategy 5	Control strategy 5

Hazard band	Exposure potential liquids			
	EPL4	EPL3	EPL2	EPL1
A	Control strategy 2	Control strategy 1	Control strategy 1	Control strategy 1
B	Control strategy 3	Control strategy 2	Control strategy 2	Control strategy 1
C	Control strategy 3	Control strategy 3	Control strategy 3	Control strategy 2
D	Control strategy 4	Control strategy 3	Control strategy 3	Control strategy 3
E	Control strategy 4	Control strategy 4	Control strategy 4	Control strategy 3
F	Control strategy 5	Control strategy 5	Control strategy 5	Control strategy 5

Table 5   Defining the control strategy level

Level	Containment measure
Control Strategy 1	No special engineering controls, with adequate control effected by general ventilation and good manufacturing practice.
Control Strategy 2	Localized capture of airborne contaminants
Control Strategy 3	Isolation of the contaminants using physical barriers between operators and hazardous materials.
Control Strategy 4	Isolation of operators from the process via direct connections, or transfer, between the process vessels and containers.
Special Control 5	Total isolation by special methods: Sealed process operating in unmanned facility with remote process control.

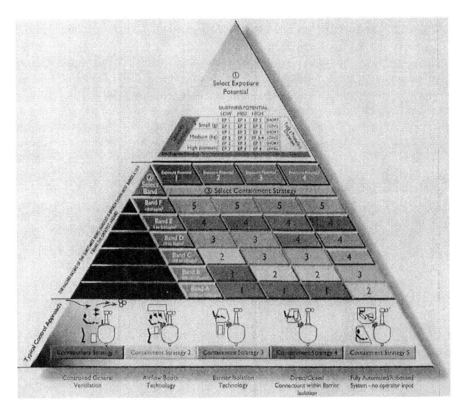

Figure 2   The control strategy pyramid.

V. UNDERSTANDING THE CONTAINMENT STRATEGY PYRAMID

When the exposure potential is combined with the material hazard band, a control strategy can be identified. Control strategies are separated into five strategy levels. Table 4 shows matrices for finding a control strategy level, first for solids, and second for liquids. Table 5 defines the control strategy levels along the range of containment measures. The selection grid forming the heart of the control strategy pyramid (Fig. 2) permits the exposure potential rating and operator exposure band to intersect at a recommended control strategy selection. A more detailed look at control strategies is presented in Table 6 for reference as you work through the control strategy pyramid for specific operations.

VI. CONCLUSION

Decisions about containment approaches within pharmaceutical companies are generally made on the basis of experience: Industrial hygienists and containment engineers are aware of myriad factors that affect emissions—where along their production lines they'll need to button up or where local exhaust is adequate to capture emissions at low levels or emissions of less hazardous products. Further, the decision makers are becoming more aware of the sophisticated and diverse equipment being marketed to control or contain at all levels of hazard. Yet, as knowledge and technology proliferate, such a cohesive framework for making containment choices as presented here can be helpful. The checklists presented in this chapter that lead to a matrix of feasible choices will of course be adjusted to reflect a unique set of production conditions, but this chapter should set a clear direction for developing a comprehensive plan.

Table 6 Control Strategy Level Criteria in Detail

CONTROL STRATEGY 2	CONTROL STRATEGY 3	CONTROL STRATEGY 4	CONTROL STRATEGY 5
ACCESS			
Only authorized personnel should be allowed into working area. Training must be provided on decontamination prior to leaving the work area	Entry to working area should be controlled. Work areas and equipment should be clearly posted. Special training on emergency evacuation and rescue must be provided	Entry to working area must be controlled. Work areas and equipment must be clearly posted. Only operators trained in emergency evacuation procedures will be allowed access to the area.	No entry to the process area will be permitted. All operator input to the process will be via remote means.
AIR MOVEMENT CONTROL	ISOLATED OPERATOR CONTROL	ISOLATED PROCESS CONTROL	TOTALLY ISOLATED PROCESS
Local exhaust ventilation should be applied at source to capture contaminants. This must be carefully designed and sited so that its performance is not compromised by external features such as draughts from doors or the building's general ventilation system. While closed systems are not always required, the efficiency of the local exhaust ventilation will be improved by partially enclosing the source (for example a fume cupboard or booth), and this degree of containment may be necessary for some applications.	Totally enclosed plant and equipment is required and must be of a standard normally encountered in an industrial environment. Only limited breaching of containment, e.g., the taking of samples is permitted. Enclosures should be maintained under negative pressure to prevent leakage. Contaminated air from the extraction system must be passed through a suitable safe change HEPA filter before exhausting outside the building. Operator manipulates compounds via glove box interface.	Totally contained process operation is required. This will be of a specialist design. No breaching of containment in operation will be permitted. Enclosures should be maintained under negative pressure to prevent leakage. Contaminated air from the extraction system should be passed through a suitable safe change HEPA filter before exhausting outside the building. Regular certification and testing of the filtration system will be required. Operator may prepare containers for transfer direct from container to vessel.	(Robotics, etc.) These designs will typically be of multiple layer containment, which uses, for example, a totally sealed process with fully welded pipe connections operating within a sealed plant envelope. Equipment design will be of a specialist design. No operator interface permitted.

MAINTENANCE AND CLEANING

Surface finishes must be easy to clean and non-porous. A regular maintenance and cleaning schedule for equipment and surfaces should be implemented. A good standard of housekeeping is expected. Cleaning should be by vacuum or wet mopping. Dry brush sweeping and compressed air cleaning should be avoided. PPR (including suitable ROPE) is likely to be required when equipment is opened for maintenance.	Surface finishes should be crevice free & ground smooth to effect easy cleaning. A regular maintenance and cleaning schedule for equipment and surfaces should be implemented. Equipment design should facilitate easy maintenance. Special procedures, such as purging or cleaning procedures such as CIP will be required before systems are opened. Permit to work systems should be considered for maintenance activities. PPE (including suitable RPE).	The highest standard of surface finish is required. This should be compatible with automated cleaning such as CIP systems. A regular maintenance and cleaning schedule for equipment and surfaces must be implemented. Equipment design must facilitate easy maintenance. Special procedures, such as purging or cleaning procedures such as CIP will be required before systems are opened. Permit to work systems must be considered for maintenance	Routine cleaning of the fully sealed process plant will not usually be required. Automated decontamination of the process will be necessary prior to any entry into the process area. Full PPE may be required even after decontamination.

PERSONAL PROTECTIVE EQUIPMENT

Respiratory protective equipment is not normally required, but there may be occasional short term activities where its use is needed. PPE such as protective overalls, gloves and goggles will be needed for maintenance procedures.	Respiratory protective equipment is not required. There may be specific short term activities where RPE or high efficiency RPE will be needed. Special PPE will be required for equipment breakdown and maintenance procedures.	Respiratory protective equipment is not required. Under emergency conditions, high efficiency RPE will be needed. Impervious overalls, gloves and eye protection should be worn.	Pressurized airsuits may be necessary for emergency conditions. Not required

TRAINING

In addition to basic induction training, specific training is required on the hazardous nature of the substances handled and the operation of the controls. Particular attention should be given to how to detect and respond to a failure in control.	Specific on-the-job training is required. This should include an understanding of the plant, the maintenance and use of PPE and procedures to detect and deal with loss of containment. Periodic retraining/refresher training will be required.	Emergency situation training is required. This should include an understanding of the plant, the maintenance and use of PPE and procedures to detect and deal with loss of containment (emergency). Periodic retraining/refresher training will be required.	Operator training and emergency situation management procedures must be set up and reviewed on a regular basis. Full liaison with the HSE at the planning stage is recommended.

12

An Architectural/Engineering Firm's Perspective

Walter W. Czander
Lockwood Greene, Augustus, Georgia

EDITOR'S NOTE

This chapter is written mainly to benefit two key audience groups: architectural/engineering firms and their clients. It's hoped that the clients of A/E firms will gain an understanding of how to optimize their use of the A/E to deliver the most valuable product possible. As well, it's intended that employees in the A/E industry will better understand their client's needs, and even on occasion be in a position to save the clients from themselves.

The client/engineer relationship can be a very rewarding one for all parties concerned. And, as with most rewarding experiences, it can also be a challenging one. We explore aspects of each, relating especially to the area of *potent compound containment*, a topic that has surfaced with renewed pharmaceutical emphasis in recent times and has every indication of continuing to do so into the foreseeable future.

I. THE EVER-CHANGING FORECAST

All processes and building designs are based upon a forecast. The forecast is usually given in packaging units per year or sales dollars of each product per year and sometimes batches or lots per year. It is safe to say there has not been a project in which the forecast has remained unchanged within even the first 30 days of that project's design phase, and in many cases more than twice during the design and construction phases of the project. A forecast for the project is developed as part of the justification for the capital appropriation that occurs anytime from two to six months (sometimes longer) prior to the engineering team actually starting work on the project. Also when new or emerging products are involved, the scheduled date for the market release is highly volatile, and this date typically gets moved up to shorten the project time. The reason for this is that the product is no longer on the back burner and it has become highly visible to Marketing, which, of course, wants to beat the competition to market, meet sales goals, or satisfy other external drivers.

In some cases, these changing forecasts decrease so much that in the client's eyes the need for containment systems cannot be cost justified. Taking one recent project as an example, the forecasted volume dropped from 30 batches a day to 2 batches per month. This change came after two months of design in developing specifications for, in this case, glove boxes and half suits to be used in weighing the material, designing containers to be used with split valves, and a drum packaging operation that was to be used in conjunction with a special shipping container. In the wake of the forecasting shift, the entire operation simply reverted to using complete protective suits, with safety packs and hoods, and loading material into conventional plastic-lined pails. Of course, this approach wound up being fairly expensive, once the invoices were received for two months of engineering that was rendered academic by changing plans.

How should a project team handle these changes?

1. Recognize that there will be a forecast change along with other changes.
2. Be flexible in developing the design. Don't always go down one path in that design. Consider the balance between operating cost, capital cost, and schedule. As an example, the following sections are a comparison of two different approaches to designing an operation to meet an emissions concentration design level of 1 to 20 $\mu g/m^3$ OEL.

II. A COMPARISON OF COSTS FOR DIFFERENT METHODS OF ACHIEVING CONTAINMENT

A. General Operational and Facility Considerations

The building design and construction section of the FDA Federal Regulations Section 211.42 requires there be separate or defined areas of operation to prevent contamination. The interpretation of this section by the leading pharmaceutical companies has led to designing facilities that have HEPA filtered air wherever the product is exposed. Also, airflow rates of 90–100 ft./min. at one foot above the operating surface and once-through air circulation are often utilized. In addition, the temperature parameters are generally designed to ensure employee comfort at 68°F $\pm$ 2% rather than for product requirements that have a wider range in temperature. All of these criteria translate into high installation and operating costs.

Additional costs in constructing and maintaining a clean room environment include special finishes, such as epoxy terrazzo floors; space needs for concealing pipes, conduits, and HVAC ductwork; and special detailing, such as flush glazing for vision panels. With all of this expense and effort, the major potential for contamination is still a factor if the area is not properly cleaned between the production run of different products, the people are not properly dressed, and/ or the people move too much through the area where the product is exposed.

When powdered material is being processed, it can be extremely important to control the relative humidity level, because humidity can cause many problems both to the product and the operators. Humidity ranges of 35–40% relative humidity (RH) are typical goals established for the product. A lower RH might be desired, but cause static electricity charges that enhance dust attraction to containers and equipment and can, in turn, hamper the collection of emitted powder by the exhaust system (if that mode of dust control is being used).

In some cases, walkable ceilings and large chases may be designed into the facility to allow maintenance personnel to perform operations such as replacement of lighting elements without entering the clean room.

Many equipment manufacturers provide equipment systems in enclosed housings, using self-diagnostics and feedback to minimize the need for direct operator interface. Still, most filling lines require some operator intervention that will expose the product to potential contamination.

The goal of applying containment technology is to eliminate the direct interface of personnel with the product and containers, thus maximizing the cleanness levels (or sterility) of the operation and at the same time protecting the worker.

B. Containment System Design

The first step in designing an operation using containment is to establish design criteria for the critical operations that will require containment. Concurrently, all processing, handling, cleaning, and maintenance operations that can affect the process must be considered and incorporated with the containment philosophy.

One way to perform this analysis, either by the client or by the A/E, is by using a detailed industrial-engineering flow diagram. (Refer to IE flow diagrams in this chapter.) Figure 1 illustrates the complete product flow and Figure 2 (an enlargement) illustrates a step in the operation that needs employee protection. The diagram is a tool for documenting every step of an operation on how material will be moved and handled. As part of the process of creating this diagram, all critical operations requiring containment can be identified. One point for both the A/E and the client: if at all possible, an experienced operator should partici-

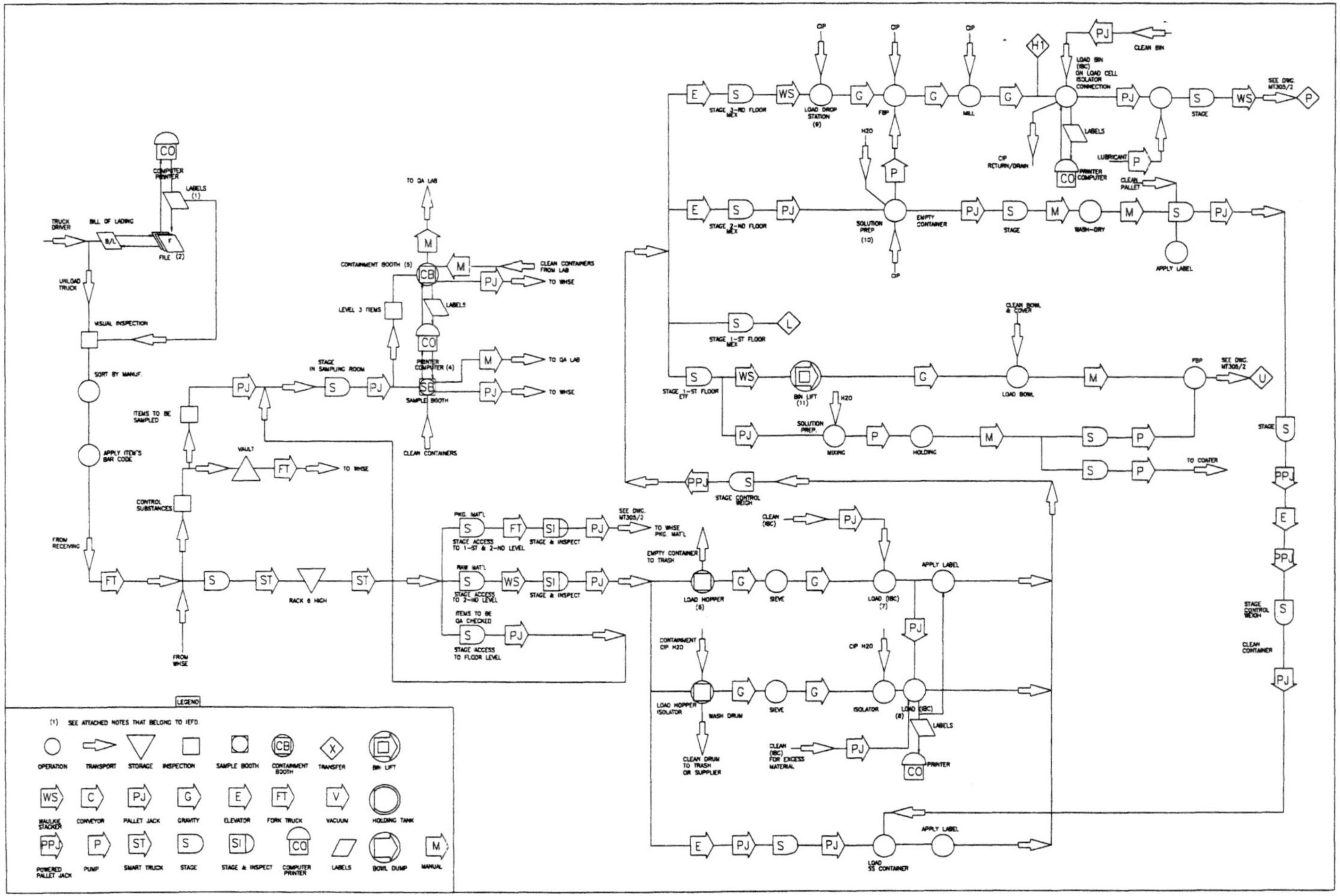

Figure 1 IE flow diagram.

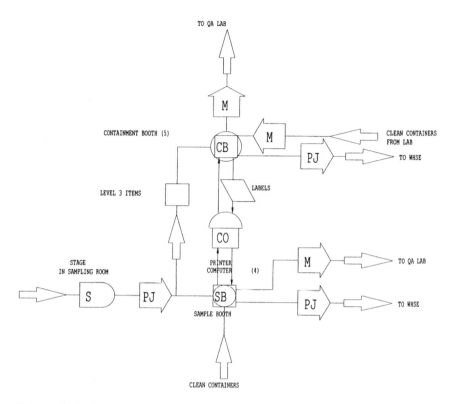

Figure 2 IE flow diagram (enlarged to illustrate areas requiring containment).

pate in this discussion. This is oftentimes not the norm in many firms, especially larger ones. However, adding this degree of realism at an early stage of the design helps to set the design onto a firmer foundation, and with it fewer changes, down the line. (Remember, down the line is when changes become much more expensive.)

Another important aspect of this analysis is the identification of material and parts transfer points, and when and where an operator must have access and could be exposed to the product. Again, having a hands-on operator involved here can be a big plus. Once the points are identified, ergonomic studies must be performed to assist in ensuring that the personnel can operate and maintain equipment in an efficient manner without violating the containment envelope.

Next, design criteria must be established for all containment units or devices (such as glove boxes, glove bags, isolators, etc.). Think of this as the "criteria for success," which the design and process will ultimately be judged against. Along these lines, clearly identify who will be responsible for establishing the design criteria, the client or the A/E. Typically the client should play a major role in this, establishing at least certain key requirements and leaving the A/E or vendor latitude to optimize the rest. Following are some of the criteria to be considered.

Batch size, which the containment unit will be housing.

Ability to replace the gloves during operation.

Need for inert gas such as nitrogen.

Sensors and controls for such parameters as oxygen and RH.

Special process material feed connections, such as those that may be required for powders or liquids.

Ability to vary external and internal illumination to prevent reflective glare back to the operator.

Access ports for container entry and exit for tools, pumps, and trash.

Pressure monitoring and control devices.

Provision for grounding to prevent discharge of static elasticity during manipulations.

Provision for fixed and temporary utilities and instrumentation connection.

Connections for air supply, exhaust, and recirculation.

Methods of sterilization, agents used, how they will be evacuated, and what, if any, effects they may have on either the machine components within the containment unit or on the containment unit itself. (Citing just one example of the amount of detail needed for a successful containment installation, be sure to check chemical compatibility of any gasketing materials with proposed cleaning agents.)

The method for performing the sterility test.

Extensive consideration of the ergonomic design of the containment unit. The designers must carefully analyze the characteristics of the equipment that is to be used by the operators. Three-dimensional simulation is one way to start, but the best proven method is to get the operators involved and to make very inexpensive mock-ups out of cardboard, thin clear plastic and wood, to prove out the design. Key ergonomic factors to consider:

1. Operators' ability to see especially when they have large gloves on
2. Glove entry versus half suits
3. Variation in operators' size and their reach limitations
4. Flexible wall versus rigid polycarbonate walls that could reflect light
5. Disassembly of equipment and how it can be removed
6. Docking methods and size of docking units
7. The removal and entry of parts, components, and material trash and scrap handling
8. Operators' fatigue in standing with their arms extended

A major advantage in utilizing containment units is that operators can work without special gowning and the building does not require a highly controlled environment, thus minimizing investment in special finishes and highly sophisticated air-handling systems. Many source-containment systems at present can be "more expensive" than the traditional approach. However, the total (net) cost to the facilities, including annual operating cost, are not that significantly different from operations utilizing the more traditional personnel-protection mode. We hope and expect that costs of source containment will continue to decrease and prove more economical with more competition and off-the-shelf design.

C. Cost Comparison

To clearly bring into focus the financial impact that the use of isolation technology can have in operation, the accompanying cost-comparison tables have been developed.

These cost comparisons are based on the building, equipment, and annual operation cost for two different projects.

1.   A sterile operation, composed of two new vial filling lines, components preparation, vial washing, and post fill area.

Table 1   General Cost Comparison of a CCRO to ITO[i]

	Initial cost		Annual cost	
	CCRO[a]	ITO	CCRO	ITO
Building	$4,431,000	$3,413,000	$177,000	$136,000[b]
Equipment	$8,800,000	$10,396,000	$1,257,000	$1,485,000[c]
Subtotal	$13,231,000	$13,809,000	$1,434,000	$1,621,000
Other costs:				
Energy			$152,000	$7,000
Gowns			$36,000[d]	$5,000
Clean-up labor			$144,000[e]	$56,000[f]
Total			$1,614,000	$1,689,000

[a] Refer to Table 2 for details.
[b] Based on a 25-year, straight-line depreciation.
[c] Based on a seven-year, straight-line depreciation.
[d] Cost is for 250 uniforms used 48 weeks/year by three operators.
[e] Uniform cost at $17.35 each × 48 weeks × 3 operators × 2 changes/week.
[f] Clean-up labor for 4 operators × 16 hours × 48 weeks × $46.88/hour.
[g] Clean-up labor for 2 operators × 12.5 hours × 48 weeks × $49.88/hour.
[h] Cost of energy, on the order of $0.10/kwH.
[i] ITO: Isolation Technology Operation; CCRO: Conventional Clean Room Operation.

2. A solid tablet operation, composed of dispensing, wet granulation microwave drying, dry milling, blending, compression, coating, and inspection.

The annual operation costs include direct labor cost, change-overs, and annual energy cost.

Some items that were not included are validation and maintenance, including replacement of HEPA filters and cleaning of aseptic corridors. Tangible and quantifiable dollar amounts for personnel safety and product protection, not surprisingly, are not reflected in the actual cost comparison.

Table 2   The Area and Equipment Cost for a Conventional Clean Room[a]

Area	Ft^2	Cost/Ft^2	Total cost
Gowning	250	$350	$87,500
Component prep	1,500	$350	$525,000
Aseptic staging	1,000	$500	$500,000
Aseptic corridor	480	$500	$240,000
Fill room 1	600	$600	$360,000
Fill room 2	600	$600	$360,000
Vial washing and post-fill			
For room 1	900	$350	$315,000
For room 2	900	$350	$315,000
Degowning	140	$350	$49,000
Mechanical[b]	6,000	$280	$1,680,000
Total			$4,431,500
Equipment cost			
Vial line 1			$2,500,000
Vial line 2			$2,500,000
Steam sterilizer			$280,000
Heat sterilizer			$290,000
Stopper washer			$500,000
Installation cost		Subtotal	$6,070,000
Equipment cost[c]			$2,730,000
		Subtotal	$8,800,000
Total costs			$13,231,500

[a] These costs are based on the average of several sterile fill and finish operations. The cost includes all fees, A/E, contractor overhead and profit, etc. It does not include validation cost.
[b] Mechanical area includes chases and HVAC equipment cost.
[c] Area cost and equipment cost are only for these areas and major equipment. A total facility would include more areas and additional equipment.

Table 3 The Area and Equipment Cost for a Containment Technology Facility and System[a]

Area	Ft^2	Cost/Ft^2	Total cost
Gowning	200	$200	$400,000
Component prep	1,500	$350	$525,000
Staging	1,000	$300	$300,000
Corridor	480	$300	$144,000
Fill room 1	600	$300	$180,000
Fill room 2	600	$300	$180,000
Vial washing and post-fill			
For room 1	900	$300	$270,000
For room 2	900	$300	$270,000
Degowning	120	$200	$24,000
Mechanical penthouse[b]	4,000	$280	$1,200,000
Total			$3,413,000

Equipment cost	
Vial line 1	$2,600,000
Glove box enclosure	$400,000
Vial line 2	$2,600,000
Glove box enclosure	$400,000
Steam sterilizer	$290,000
Isolator	$100,000
Stopper washer	$510,000
Isolator	$100,000
Filter integrity testing	$70,000
Isolator	$100,000
Installation cost	$3,226,500
Equipment cost[c]	$10,396,500
Total costs	$13,809,500

[a] These costs are based on the average cost for two recent projects using containment technology, one in New Jersey and one in Puerto Rico.

[b] Mechanical area includes HVAC equipment.

[c] Area cost and equipment cost are only for the previously mentioned areas and major equipment. A total facility would include more areas and additional equipment.

Table 4 Compares a Solid (Tablet) Operation Using a FBD with a Microwave Dryer Closed System MDCS[a]

	Initial cost		Annual cost	
	FBD	MDCS	FBD	MDCS
Building[b]	$3,805,000	$2,335,000	$152,000	$193,000[b]
Equipment	$6,496,000	$8,490,000	$927,000	$1,213,000[c]
Subtotal	$10,297,000	$10,825,000	$1,029,000	$1,306,000
Other cost				
Energy			$100,000	$85,000
Operation			$210,000	$105,000
Total			$1,389,000	$1,496,000

[a] The microwave granulation dryer has one less step, is more contained in its design and is, along with the fluid bed dryer, being used by members of the pharmaceutical manufacturing industry.

[b] Based on a 25-year, straight-line depreciation.

[c] Based on a seven-year, straight-line depreciation.

Table 1 compares conventional clean room costs with containment technology costs; Table 2 examines the area and equipment costs for a conventional clean room; Table 3 discusses the area and equipment costs for a containment technology facility and system; Table 4 compares a solid operation, using a fluid bed dryer (FBD) operation, with a microwave dryer closed system (MDCS), using a glove box to handle the potent material; Table 5 examines the area and equipment cost for the FBD operation; and Table 6 the area and estimated cost for the MDCS facility and system.

1. Summary of Cost Comparisons

The life-cycle cost analysis shows an increase in cost of the Isolation Technology Operation (ITO) over the Conventional Clean Room Operation (CCRO) in the initial investment and the annual cost. However, the truly significant advantage of containment technology is its ability to provide a higher degree of protection to the product and improved safety for employees.

As stated earlier, tangible and quantifiable dollar amounts for personnel safety and product protection, not surprisingly, are not reflected in the actual cost comparison. It is noteworthy to see that there is only a minor cost differential between the traditional and the source-containment scenarios, even with no quantitative consideration at all being given for either product quality or worker safety. As the tables demonstrate, there are a multitude of ways to analyze the costs of a facility. Beyond the raw numbers, there are other factors to consider as well.

Table 5 The Area and Equipment Cost for a FBD Operation

Building cost

Area	Ft^2	Cost/Ft^2	Total cost
First level			
Gowning, office support	1,200	$200	$240,000
Dispensing	700	$350	$245,000
Bin loading	250	$300	$75,000
Bin blending	300	$300	$90,000
Staging, aisles, stairway, etc.	1,000	$280	$280,000
Second level			
FBD controls	600	$250	$150,000
Compression (2)	700	$340	$238,000
Coating	500	$340	$170,000
Support	1,200	$300	$360,000
Staging, aisles, stairway, etc.	1,000	$280	$280,000
Third level			
High shear mixer	600	$300	$180,000
Drop station to compression (2)	500	$300	$150,000
Drop station to coating	250	$300	$75,000
Mechanical	1,150	$220	$253,000
Staging, aisles, stairway, etc.	1,000	$280	$280,000
Fourth level			
Bin unload to high shear mixer	250	$300	$75,000
Mechanical	2,000	$220	$440,000
Staging, aisles, stairway, etc.	800	$280	$224,000
Total building			$3,805,000

Equipment cost

Dispensing (bin loading)	$50,000
Bin unload station	50,000
High shear granulator	500,000
Fluid bed dryer	2,000,000
Recovery system exhaust	300,000
Mill	50,000
Platform	40,000
Chutes and conveyors	7,000
Lubrication loading station	25,000
Cleaning—CIP	450,000
(2) Compression loading stations	100,000
(2) Tablet presses	1,600,000
(10) Plastic bins	100,000
(6) IBC (including docking valves)	90,000
Bin washer	250,000
Tool room—misc.	60,000
Coating machine loading	50,000
Coating machine	670,000
Total	$6,496,000

Table 6 The Area and Equipment Cost for a MDCS

Building cost

Area	Ft^2	Cost/Ft^2	Total cost
First level			
Control room	250	$250	$62,500
Coating	500	$300	$150,000
Gowning, support	800	$250	$200,000
Staging, aisle, etc.	1,000	$220	$220,000
Second level			
Compression	350	$300	$105,000
Blender	400	$300	$120,000
Support	200	$250	$50,000
Stairway, aisles, etc.	1,000	$250	$250,000
Mechanical	600	$220	$132,000
Third level			
Microwave high shear mixer milling	500	$300	$150,000
Milling			
Stairway, aisle, etc.	600	$250	$150,000
Mechanical	1,000	$220	$220,000
Fourth level			
Dispensing	600	$300	$180,000
Glove box	100	$250	$25,000
Stairway, aisle, etc. (Process)	400	$250	$100,000
Mechanical	1,000	$220	$220,000
Total building			$2,334,500

Equipment cost

Dispensing	
Scales	15,000
Laminate flow hood	120,000
Bench	10,000
Dust control pick-up, etc.	25,000
Installation	60,000
Solution prep	
Vessels	160,000
Platform	30,000
Scales	15,000
Dust control pick-ups, etc.	25,000
Installation	100,000
Mixer	
Microwave granulator dryer	825,000
Infeed connection to granulator	25,000
Installation	245,000

Table 6 Continued

Building cost Area	Ft^2	Cost/Ft^2	Total cost
Mill rotary			65,000
Dust collector—pick-ups, etc.			35,000
Infeed to intermediate bin			20,000
Installation/intermediate bins (8) 13,000			104,000
with passive valve (8) 12,000			96,000
Compression			
Bin unload station (2) 35			70,000
Compression machine (2) 750			1,500,000
Conveying to coating hopper			40,000
Installation			450,000
Coating			
Hopper vacuum load			70,000
Coater			690,000
Coater discharge			60,000
Installation			300,000
Cleaning system CIP/COP (installation included)			450,000
Blender			
Bin			350,000
Installation			120,000
Spare parts and tools			300,000
Total equipment			$6,375,000
Engineering, construction, management, contractor			
Overhead and profit, fees, etc.			2,125,000
Total			$8,500,000

III. INEXPERIENCE OF PEOPLE

There is a saying, intelligence solves problems, experience solves them faster. The inexperienced person can be the prey of the vendor's snags. Vendors can imply promises that an experienced person can immediately challenge or at least clarify. As an example, when a vendor gives a schedule for delivery, he might typically say something along the lines of "It will take 16 weeks for delivery of equipment." What is really being said, quite appropriately, is that delivery will be 16 weeks after shop drawings are approved. Since the shop drawings are sent to the client until up to four weeks after the client issues a purchase order, and since the client's own organization, if true to form, can take another four weeks for the checking and turn around of the drawings, then once time is added in for changes (requiring resubmittal of the drawings), the schedule is now closer to 30 weeks instead of the originally stated 16 weeks.

The preceding example lends itself especially well to containment design. Containment design oftentimes requires containment devices (glove boxes, glove bags, isolators, etc.) or the modification or customization of the direct production equipment. These pieces of "containment equipment," however, are precisely the elements that are the most dependent on shop drawing approvals to begin with.

It is the responsibility of the A/E consultant to bring to everyone's attention the notion of caveat emptor, or buyer beware. To be sure there is no confusion, the vendors must be asked to quote only on the exactness of the specifications, and not to have embedded in the quotation references to additions or deviations that "would be of benefit to the project." That said, however, the door should be open for them to add supplemental ideas, with additional cost and time consequences stated along with claimed benefits to the project, with clear demarcation from the base quotation.

IV. WAITING FOR THE CLIENT TO IDENTIFY THE OEL REQUIRED AT EACH STEP OF THE OPERATION

Typically, the client's Life Safety Engineer, Industrial Hygienist, QA Department, or Project Manager has the responsibility to establish an acceptable level of exposure. This level is usually based on the active ingredient being a highly potent, pharmacologically active or toxic agent, along the lines of pathogenic materials, cytotoxins, synthetic hormones, transgenic microorganisms, or other compounds of similar issue. The guidelines for operators' exposure for these type products are in most cases issued by the Advisory Committee on Dangerous Chemical in the United Kingdom and by the Centers for Disease Control in the United States in conjunction with the National Institutes of Health (NIH) and the European Economic Community (EEC).

The initial design, therefore, starts out to be conservative in allowing zero risk to the operator and the environment. On past projects, containment design criteria have changed, through either the amount of active ingredient becoming diluted or the amount of dust emitted during the operation.

Again, the team should be alert to the fact that this has a good chance of happening, and should consider alternate designs as the project continues so it will be prepared to quickly make necessary changes in the need for less containment such as reduction in the room sizes or elimination of air locks.

V. ACCEPTANCE OF THE DESIGN

This should be straightforward, especially when there is an off-the-shelf design that can be used. Unfortunately, it can be one of the most difficult things to do.

Here are some of the don't's and do's.

1. Don't design the containment system by yourself. Do the design with the team and the user even if it means you have to compromise what you may consider to be the ''best'' design. After all, the user is the most important group; they are the ultimate clients.
2. Don't think that everybody can understand, much less fully comprehend, the design drawings. Relatively few people can translate even three-dimensional drawings into a full-scale operation. Make a full-scale mock-up (inexpensive materials are fine) as exact as possible to the proposed finished containment unit.
3. Don't forget cost or schedule. Do continually remind all involved that there is a schedule to be maintained and a budget to be met.

VI. MEETING THE PROJECT SCHEDULE

When the project starts, the people that are initially involved are made to understand the importance of meeting the schedule. However, as the project moves along, more people become involved, and many of the original group are no longer on the project. The new people are interested in completing their assignment and are not as cognizant of the overall design schedule, much less how their effort effects that schedule. This is true within both the A/E and the client organizations.

One of the best ways to keep control over the project and the schedule is to make sure everyone who starts to work in the project understands the project goals along with the schedule in specific days and hours. Then get a commitment and an agreement in writing from them as to how many hours they will be spending, along with a completion date for their effort.

VII. MINIMIZING CHANGE ORDERS AND PROJECT OVERRUNS

Why does this happen, even when the ''best people'' are on the job? After you recognize and analyze the reasons, you should be able to better prevent it from happening again. Here are some causes:

A. Poor Scope of Work

The scope of work starts with a request for a proposal given to the design firm to bid on, along with the time to respond (typically about two weeks). The time given for all projects seems to be this same length, whether for a small feasibility study or a 50-million-dollar project. Obviously, the A/E will provide the best bid possible given two weeks, or whatever other time frame the client requires. However, the client should be aware that an unrealistic time for the design team to respond will lead to quick decisions and increased assumptions, and that the design team will hope to rectify any inaccuracies when the project starts. Unfortunately doing so after the fact can be highly unlikely, as it usually takes several additional weeks before the project "starts," and those items that initially needed clarification (and still do) are not put on the table until it's too late. When these items are brought to the attention of the client's Project Manager, it generally winds up adding more time for clarification and, of course, more unforeseen dollars.

B. Changes in Personnel

Any change in personnel, whether in the design firm's team or a member of the client's team, has an impact. There is a learning curve to be considered for all people working on the project, along with all the items and ideas that were hashed over and discarded, along with the reasons that a concept was selected. Building on this point a bit more, the one who has the biggest impact on changing the direction and project cost is the client's decision maker. And when that person is changed it can add weeks to the project. Because time is money, it can have a major impact on the project cost as well. What can hurt even more is when the client's decision maker is not closely involved and has a subordinate as a design team representative who is not empowered to make a decision. How many times have we witnessed a presentation being made to such a person, and hear "Let's look at this from a different perspective." Sometimes that is a legitimate comment; often, however, it can be translated as, "Show me at least five other concepts, with pros and cons and weighted numbers, and maybe I'll be able to get some direction from my boss sometime soon." This person's indecision can add weeks to the project, and again, time being money, more dollars will be needed.

C. Shop Drawings: Late, Inaccurate, or Missing Details

Problems with shop drawings, for either process equipment or containment devices, can cause major containment design problems. Often, operator ergonomics and interface requirements are one of the key attributes that are designed into any piece of containment hardware. The success or failure of a "workable operator interface" can hinge on a mere couple of inches in accuracy of where certain gloves, vision ports, interlocking hardware, or other components have been located. Specifically, these locations need to be coordinated with the equipment they're adjacent to. Any inaccuracy or missing detail in drawings can impact this directly. The best way to resolve this problem is for either the A/E (typically) or the client to develop a good relationship with the equipment suppliers, visiting their operations, and making it a strong written item in the vendor bid package that the shop drawings and required utility lists are needed by a date certain.

D. Splitting Up a Project or Having too Many Different Design or Construction Firms on the Job

This can lead to several types of activity and behavior, none of it beneficial to the project. Inefficiencies can include:

Calling more meetings so that everyone is kept up to date on decisions and new issues; essentially everyone staying on the so-called same page.

Duplication of the work effort from clerks and administrative assistants, who have to first schedule the additional meetings, and then distribute the additional meeting notes to at least twice as many people (who many times then don't read them or respond in time to issues before the following meeting).

Depending on the personal chemistry of the major players involved, an unhealthy one-upsmanship can sometimes develop. In this situation one or more parties will unduly critique any and all suggestions by other parties they consider themselves to be "in competition with," rather than contributing to a harmonious team who has as its focus keeping the project on schedule and within budget.

VIII. CONCLUSION

Containment technology has been used for many years in certain areas, such as the nuclear, disease research, and asbestos-abatement industries. Only relatively recently has it been put to greater use for sterile or toxic operations in the pharmaceutical industry. More suppliers and designers are entering the field, bringing with them improved equipment, methods of equipment docking, abilities to clean equipment in place without disassembling, and improved ergonomics that allow the operator better interactions with the processes. It is important that the A/E continually stay current with the technology and terminology changes in the containment system designs, especially with the increasing number of potent drugs that are found to have major and suspect effects to the operators and the environment.

13
Containment in the Hospital Setting

M. Michele Moore
Containment Technologies Group, Inc., Indianapolis, Indiana

Containment technology in the hospital setting may mean isolation of patients to prevent the spread of contagious disease. Or it can mean separation of the caregiver from contaminated body fluids, potent medications, or other hazardous materials. The focus of this chapter is separation of personnel from potentially hazardous materials in the hospital laboratory and in the pharmacy.

I. LABORATORY

The most prevalent protection in the hospital laboratory is the Class II biological safety cabinet, which does offer some level of protection for the integrity of the tests. But the biological safety cabinet can fall short of providing proper levels of protection for the individuals performing the testing, based on information coming from the pharmaceutical industry. Simple observation of a typical Class II cabinet shows that surfaces cannot be cleaned or disinfected. Its noncoved corners create difficult to reach areas, and a large filter face is generally protected with a cover that is difficult to clean or disinfect.

A typical routine of operations in the laboratory includes receiving samples, preparing samples for testing with a variety of analytical methods, and disposing of the sample materials. For some laboratories, this is a high-volume activity involving hundreds of samples per day. This sequence of activities has a number of areas in which personnel protection could be improved.

II. SAMPLE RECEIVING

Personnel protection may need improvement beginning with the first step of sample receipt. Container exteriors may be contaminated, and in many cases the containers are delivered to the laboratory and placed in a holding area with no decontamination procedure. The container, therefore, can be a first major source for contamination.

Proper gloving provides protection from dermal transfer of contaminant but does not protect the individual against inhalation risks. To best protect personnel receiving samples, methods need to be improved to prevent container contamination. A designated area with a controlled environment should be established where the specimens can be delivered. An ideal entry point is a pass-through enclosure with glove ports on the laboratory side that allow receiving personnel to perform a decontamination routine before introducing the samples into the laboratory. The enclosure should have a filtered air source that filters both inlet and exiting air.

III. SAMPLE PREPARATION

Sample preparation involves transfer of material to be tested, whether fluid or solid, to the proper container, addition of materials to provide the proper media, and placement of test materials into the container for receipt by the test instrument. Potential exposure to contaminated materials at this point is extremely high and in many cases is exacerbated by such support materials as syringes and sharp objects, which can puncture gloves.

Two potential hazards to personnel working in this environment should be addressed. The first is dermal exposure and the second is exposure by inhalation. Dermal exposure is a visually apparent instance of test materials coming in contact with exposed skin. Personnel protective devices such as gloves, masks, and apparel can minimize this route of exposure. The second exposure potential is that of aerosols or airborne liquids, which can enter the body more directly through the nose, mouth, or eyes. These routes provide direct absorption and, therefore, a greater danger.

A large percentage of samples are prepared in the open laboratory environment and can present a danger both to individuals preparing the sample and to their fellow workers. Some samples are prepared in a laminar flow environment, which may or may not be a biological safety cabinet. Where laminar flow is used to protect the sample, a horizontal flow unit could conceivably be sending contamination throughout the entire facility. Biological safety cabinets offer adequate protection for some of the test materials, but other materials require the level of protection found only with barrier or isolation technology.

IV. TESTING

Testing equipment can create airborne material where pumping or transfer of the material occurs under pressure. For highly contagious or extremely high-risk fluids that could contain hepatitis or HIV virus, for example, both the test protocol and the equipment should be closely evaluated for contamination potential.

A. Discussion

Tracking of disease rates and adverse effects from potential exposures has not been commensurate with data collection and analysis taking place in other life sciences industries such as pharmaceuticals. Without substantiation of adverse effects, it is dangerously easy to ignore problems.

The Center for Disease Control has identified the major issue of increased infection rates and exposures to contagious disease, but has not taken the analysis to the next level of segmenting the major hospital activities to determine the areas that pose the largest risk to patients and employees. The five areas within the hospital that are the most likely candidates for exposure potential are the patient room, surgery suites, emergency areas, laboratories, and the pharmacy.

V. PHARMACY

The hospital pharmacy is responsible for dispensing parenteral products. Many of these products are purchased by the hospital in a concentrated form, usually in vials or ampoules, and require dilution before administration. Pharmacists and technicians working in the IV area prepare the products for the patient by adding to the pharmaceutical what is referred to as a "piggyback" carrier of sterile dextrose in water or saline before infusion at the patient bedside.

Regulations concerning preparation of parenteral products in the pharmacy are developed and enforced on a state-by-state basis rather than by a single federal agency such as the FDA. This has resulted in a wide variation in rules governing facilities. In 1998, New Jersey was the first state to enact rules requiring hospital pharmacies to prepare "sterile" products in a class 100 environment located in a class 1,000 clean room with a class 10,000 anteroom or with barrier isolator technology. This was the first major improvement in sterility assurance level for the hospital pharmacies since laminar flow technology was introduced.

Most states require that parenteral products be prepared in a class 100 environment, which typically employs laminar flow, a technology that has been used in the pharmacy for over thirty years. A significant number of the laminar flow installations are horizontal benches that provide some product protection but leave the personnel preparing the product open to potential exposure to antibiotics and other potent pharmaceuticals. The laminar flow technology used for product protection can for some compounds be seriously wanting as a measure of personnel protection and can, in fact, also allow cross-contamination within the entire pharmacy. Until recently, when excessive antibiotic exposure was identified as a health problem, this was not a concern in hospital pharmacy.

As late as 1995, standard practices published by societies supporting the hospital pharmacy did not recommend the use of gloves but to "scrub hands and arms with an appropriate antimicrobial skin cleaner" before performing aseptic manipulations with products for immediate administration to patients.

Concerns and studies about exposure to cytotoxic agents in the early 1980s resulted in recommendations that this class of compound be prepared in class II biological safety cabinets or vertical laminar flow hoods. Data published in 1999 indicated that significant surface contamination was detected in six hospital cancer centers using the class II cabinets (1). This confirms pharmaceutical information that laminar flow technologies, while very useful, have limited protection capabilities (2).

A. Discussion

Containment technology in the hospital pharmacy requires improvement. The introduction of barrier isolation technology by regulation in a limited number of states is the first major step to improve personnel protection since the introduction of vertical flow hoods. This technology will likely gain more acceptance as the limitation of the vertical airflow hoods is communicated.

Barrier isolation technology offers two important protections: the first is protection of personnel from the product and the second is protection of product from personnel. Personnel protection from the product is critical, because pharmacy personnel routinely find themselves in potential exposure situations involving potent pharmaceuticals that may cause adverse reaction. Protection of the product from personnel is equally important, because the recipient of the end product is likely to be in a compromised state of health and more likely to develop an infectious disease or other physiological reaction.

The Center for Disease Control reported that nosocomial infections in hospitals have risen dramatically in the United States (3). Statistics show that hospital-acquired infections result in the loss of 90,000 lives and at a cost of $4.5 billion per year. Many of these losses may be caused by lapses in aseptic technique while handling medications that support rapid bacterial growth (4). Hepatitis and infections of the human immunodeficiency virus and antibiotic resistant bacteria can go undetected by the carrier for extended periods of time.

For a carrier such as a health-care provider working in the pharmacy, the risk of transference in an open laminar flow system compared with a barrier isolation system is greater by several orders of magnitude, because the physical separation between the individual and the product is only a curtain of air.

Figure 1 Ergonomic workstation.

VI. CONCLUSION

Containment technology in the hospital laboratory and pharmacy environment should be upgraded to reflect current data and technology. Barrier isolation technology is just beginning to be recognized as a means of providing a higher level of protection for products, samples, and personnel. As this technology is integrated into the hospital segment of health care, it will be important to design systems that offer protection both to product and personnel. Exhaust air from barrier isolators used in these settings must be properly filtered. Figure 1 illustrates one example of a currently marketed workstation incorporating proper ergonomic design and flexibility.

REFERENCES

1. TH Connor, RW Anderson, PJM Sessink, L Boadfield, and LA Powers, Surface Contamination with Antineoplastic Agents in Six Cancer Centers, American Journal of Health-System Pharmacy, July 1999, 56 (14).
2. Hierarchy of Containment Technologies, Presentation, International Society of Pharmaceutical Engineering, Containment Conference, Washington, DC, June 1999.
3. Health Facilities Management Magazine, May 1998, 11 (5).
4. K Kimsey, Nosocomial Infections—United States, Health Letter on the CDC, September 4, 1995.

Index